ROBOT INTELLIGENCE
... with experiments

Other TAB Books by the author:

Dedication:

This book is dedicated to experimenters who lack the appropriate academic credentials, but qualify as true scientists by virtue of their imagination, talent, dedication to hard work, and proven performance.

No. 1191
$16.95

ROBOT
INTELLIGENCE
... with experiments
BY DAVID L. HEISERMAN

TAB BOOKS Inc.
BLUE RIDGE SUMMIT, PA. 17214

FIRST EDITION

FIRST PRINTING

JANUARY—1981

Copyright © 1981 by TAB BOOKS Inc.

Printed in the United States of America

Library of Congress Cataloging in Publication Data

Heiserman, David L 1940-
    Robot intelligence . . . with experiments.

    Includes index.
    1. Artificial intelligence—Problems, exercises,
etc. I. Title
Q335.H44   001.53'5   80-21440
ISBN 0-8306-9685-7
ISBN 0-8306-1191-6 (pbk.)

Cover art courtesy of Computer Graphics, Melvin L. Prueitt, Dover Publications Inc.

Contents

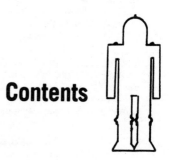

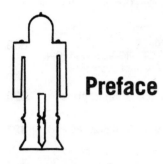

Preface

The robots are coming. There can be no doubt about that. The only questions are: How? When? and Who?

I am convinced that this book represents a step in the right direction for answering all three questions. It shows *how* the job can be done, it suggests *how long* it will take, and it invites *you* to participate in a meaningful way.

Above all, this is a book for people who want to work with machine intelligence and the germ of real robots on a first-hand basis. It's one thing to read a book outlining the history of robotics, summarizing a lot of on-going activity in the field and suggesting some exciting prospects for the future. It is quite another thing, however, to reach out and grasp the whole matter with your own talent and imagination.

This is a book that presents a particularly vigorous and rich form of machine intelligence in the framework of computer simulations. That isn't to say that the intelligence, itself, is simulated. Quite the contrary. The only thing that is simulated is the mechanical action of the robot machine which employs the intelligence.

While there are indeed some distinct advantages to fitting machine intelligence into an actual working model of a robot (see *How To Build Your Own Self-Programming Robt*, TAB book No. 1241) there are also some advantages to simulating the behavior on the screen of a home computer system. In fact, the two approaches complement one another quite nicely.

Are you ready to go to work? Good. I'm looking forward to reading about your results in some technical magazines. Who knows? Maybe I'll even be seeing you on TV!

David L.Heiserman

A Fresh Look at Robotics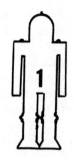

On the evening of January 25, 1921, Karel Capek's play, *R.U.R.* (Rossum's Universal Robots), gave birth to the 20th Century notion of artificial men and introduced the word *robot*. The birthplace was Prague, Czechoslovakia, but in less than two years, audiences in Germany, England, and the United States were applauding German and English versions.

The times were right for such a play, and its influence remains with us today. Unfortunately, the influence of *R.U.R.* is not an entirely positive one.

R.U.R. is responsible for some serious misconceptions about robotics that have yet to be ironed out. The central theme of the play is that technology can lead humanity along a road to total destruction. What begins as an exciting and beneficial technological enterprise can, if left unchecked, get out of hand very suddenly.

And indeed, there are many unenlightened people today who truly fear the coming of real robots and have every serious hope that robots will remain forever only on the pages of science fiction books and in the imaginations of film and TV producers. These people aren't "fooled" by the positive "robot propaganda" of the film, *Star Wars*, nor are they comforted in the least by Isaac Asimov's Laws of Robotics.

If you ever become involved in modern robot work, even as an amateur experimenter, you will eventually feel some pressure from people who do not agree you are doing anything positive for the welfare of humanity.

Serious social resistance to the development of robots will not become a big problem for some time, however. It is a problem we will have to reckon with one of these days, but it isn't an immediate concern.

9

There is a second feature of *R.U.R.* that does pose a problem of immediate concern. The word *robot* is derived from a Czech expression *robota* which implies a quality of slavery or servitude.

Nowadays human slavery is regarded as a very nasty thing in most of the world. On the other hand, most people agree it would be quite nice to have a slave or two to take care of our more mundane and difficult chores. Since human slavery is out of vogue, the convenient alternative is a mechanical man that is capable of doing the same work as a human slave, but without bringing us face to face with any serious moral issues.

We thus face a rather arbitrary, but firmly established doctrine—The whole purpose of robotics is to sidestep the moral issue of human slavery and come up with a machine that does our bidding with great energy, competence, and unquestionable loyalty.

"Ah, how nice it is that you are building a robot. What does it do for you?" Does it answer the door and telephone for you? Does it sweep the floor or carry out the garbage? Does it mix drinks and fetch the evening paper?" In other words, "How good is your *slave machine*?"

Such a banal attitude toward the purpose of robotics was engendered, at least in our historical era, by *R.U.R.*, and it has been perpetuated by most robot fiction as well as more serious robot research.

The notion that a robot ought to *do something* in response to its human master's commands is threaded through virtually all thinking about robots. To see how well the notion has served the cause of robotics, just look around at all the nifty robots serving us today—there aren't any, are there?

So *R.U.R.* is really the foundation for most robot thinking today. Take from that play its social commentary and the servile implications of the word *robota*, and you have a positive vision of what robots ought to be. But, look only at the social commentary and the implications of *robota*—as most people do— and you have a couple of serious problems to contend with. *R.U.R.* thus sparks the vision of robotics, but at the same time, draws some lines that few researchers have dared to cross or, indeed, realize can be crossed.

Of course, *R.U.R.* doesn't epitomize *all* the problems of modern robot research. WWII provided the motivation for doing a lot of research in electronics and machine control, and post-war researchers found they had a whole inventory of new devices and theories to make their work more exciting and meaningful.

Physiologists jumped onto the electronics bandwagon quite early in the game, first using new and sensitive electronic equipment to measure physiological activity of an electrical nature (in nerve, muscle and cardiac tissues) and then to build electronic analogues of biological systems.

Medical journals suddenly began looking something like *Popular Electronics*. To the untrained eye, it was difficult to tell the difference between a circuit for a one-tube radio receiver (usually built into a piepan or a wooden cheese box) and an electronic model of a brain cell.

The unfortunate side of the matter, as far as robotics is concerned, is that researchers seemed to lose sight of the fact that the electronic models of physiological systems were based on biological theories. Most of those theories have undergone a number of revisions through the past 25 years, and the most successful medical researchers rarely resort to electronic modelling. Biochemistry is the thing these days, but a lot of robot researchers don't seem to recognize that fact.

The nifty electronic analogues of biological systems developed some 25 years ago, and since chucked as useless novelties by the physiologists, remain as the basis of far too much robot work today. To be sure, the circuitry has been updated to include computers and microprocessors, the roboticists aren't altogether ignorant of the most recent views of physiology. The problem is an insistence upon employing an outmoded procedure of robotics by analogy.

The thinking goes something like this—if we want to build a machine that mimicks the behavior of a man (*i.e.* a robot), we must begin by building electronic systems that mimick the activity of human/animal parts.

It is a case of analogy piled upon analogy. Given the time and funding, one can construct some highly elaborate analogies of animal organisms. Given the proper training in physiology and electronics, an ingenious (and well funded) researcher can come up with a pretty nifty "eye" that can recognize certain elementary patterns and distinguish one kind of motion from another.

It's possible to build a "finger," too. It can sense pressure and temperature, and respond by moving around on the workbench. Such electromechanical analogues have been around for quite some time now, and most work along this line concerns applying modern microelectronics to get the physical size down to reasonable proportions or extend the range of possible activity.

This is not to say that researchers are wholly ignorant of the limitations of building analogies of biological systems. On the contrary. We all realize that a working analogue of a simple animal brain would require vast compute facilities now beyond the reach of economic reason.

Analogue builders are left in the position of treading water until further developments reduce the size, increase the sophistication, and lower the costs of the parts. The general attitude is this— we will do the best we can with what we have right now. And when the time is right (that is to say, when some sort of great breakthrough occurs), we will be ready for it.

Analogue builders, nor the most part, adhere to an admirable faith in technology. It is a sort of messianic faith that someone, somewhere, sometime will come up with a discovery that will let us put together all the bits and pieces of physiological analogues into a working unit of manageable physical proportions.

The possibility that there is an alternative to building analogues is practically unthinkable. At best, such as notion is something like contemplating the sound of one hand clapping—it makes sense only in the most obtuse philosophical framework. At worst, it smacks of heresy.

Nevertheless, this book suggests an alternative approach to developing real robots. It suggests that the notion of building analogies of physiological systems ought to be abandoned in favor of an alternative that promises to do away with the need for counting on some great gift of future technology.

The subject of this book, evolutionary adaptive machine intelligence, brings those "future developments" right up to our doorstep. That is not to say you will be able to realize the robot of your dreams overnight, or even within the next five years or so. A new idea of this sort requires a lot of sound and responsible research before it is ready for full-scale implementation. The experiments outlined in this book represent the very beginning of the job, there is much more work to be done. But, the tone of the research is entirely different from that of traditional analogue builders. In our case here, the main hurdle is the sheer bulk of the work that had to be done. The nature of the work is quite well defined and well within the grasp of current technology, however. The idea is a workable one, and all this work concerns refining the idea and proving that it is a viable one.

Besides marking a departure from conventional analogue thinking, the ideas about robotics presented in this book call for doing away with the *R.U.R.* notion that a robot must be a servile creature. Viewing a robot as a mechanical creature with some aura of independence is not wholly necessary to the work that must be done at the present time, but it sets the stage for an attitude that narrows the work into more productive channels.

What does this new approach offer in terms of allaying fears about robotics? Nothing that I can see right now. In fact, some of the programs listed in this book create mere spots of light on a TV-like screen that behave in a fashion sometimes bordering on being awesome. That spot of light cannot hurt anyone, of course; but, if you understand exactly what that spot of light is doing and the mechanisms behind its behavior, I can promise there will be moments when you wonder whether or not you are playing with something that might get out of hand when its transformed into a real creature.

Does the prospect of building an intelligence without human control bother you? Does it excite you? Does it both bother and excite you? Fear comes from ignorance and excitement comes from engaging in activities that feed excitement. In any event then, you ought to do the work suggested here.

A BRIEF AND PERSONAL HISTORY OF EAMI

Take a moment, if you will, to consider a brief and somewhat personal history of evolutionary adaptive machine intelligence. The purpose of this little dissertation is twofold—first, to show that the idea did not come about overnight, and second, to set the stage for the more technical and theoretical discussions of the matter.

I began thinking about robots seriously in 1974. At that time, I followed the sort of thinking that typifies that of many amateur roboticists today. My immediate goal was to build a machine that would obey my commands and yet include some automatic features that would allow the machine to operate somewhat independent of direct, literal commands.

The result was a charming little machine I called *Buster*. Complete plans for a working model of Buster can be found in my book, *Build Your Own Working Robot*, TAB book No. 841.

While Buster eventually satisfied my original goals and, indeed, exceeded them in some respects, I felt a certain kind of dissatisfaction. Buster was OK, but there seemed to be a necessary next step to be taken.

Building Buster was an exciting job, and the more I worked on the manuscript for the book, the more excited I became. I think that element of excitement is reflected in that work. But, the excitement was not well defined.

Setting Buster aside, I began searching for some definition of the strange combination of dissatisfaction and excitement. Looking around at what others were doing with robots—both on the amateur and professional levels—I began seeing the people running headlong into the same barrier. That barrier was the exceeding complexity of the systems they envisioned. It is one thing to build a little machine that creeps around a room, avoids obstacles and understands a small vocabulary of spoken words. That's all nice, fun and a necessary part of robot work. However, to extrapolate the ideas and mechanisms to the kinds of machines people want robots to be posed some serious problems.

There is a rather practical principle of science that says something like this—if a theory leads to greater complication than simplicity, its probably wrong. Oh sure, I know about the idea that every new theory poses more questions than it answers, but that isn't the point here. Any theory that is worth anything tends to simplify the matters it deals with. Einstein's Theories of Relativity certainly pose more challenges, but at the same time they tidy up our view of the universe as compared to the older notions of Newton.

In 1976, it wasn't difficult to see that robotics was getting itself into a mess. Using the general approaches of that time, things were getting far too complicated. It was time to come up with a new theory that would pull everything back together again.

Two magazine articles written during that formative period reflect my attempts to devise a new and more general theory of robotics. "What Is a Robot?" (*Personal Computing*, July/August, 1977) defines what a robot should be and what it should not be. The second article, "Good Grief! Now there's Rodney and Buster" (*Modern Electronics*, March, 1978), suggests a four-step evolutionary approach to the development of intelligent robots.

While writing that second magazine article, I was in the middle of a new robot project. The new machine, *Rodney*, was something entirely different from Buster. Rodney reflects the first three stages of evolutionary robotics, and one of the more obvious results is a mechanical creature that tends to exhibit personality traits.

Rodney supports the workability of the evolutionary approach to robotics. The project clearly demonstrates that it is possible to build a self-programming, self-directed, and independent creature, using relatively simple and commonly available parts.

The Rodney project is completely summarized in my book, *How To Build Your Own Self-Programming Robot*, TAB book No. 1241. Rodney represents a new generation of robots, using a microprocessor technology in place of the purely TTL technology of the Buster machine. That technological improvement, however, is incidental compared to the conceptual differences.

The evolutionary approach to robotics, as exemplified by Rodney, allows a machine to evolve in the context of machines, as opposed to force-fitting a machine into a biological context. A machine allowed to evolve within its own context eventually begins exhibiting forms of overt behavior that are markedly similar to living creatures. And it is at that point— at the level of overt behavior— that we can begin drawing legitimate parallels between animal and robot activity.

The Rodney machine is a successful project. While watching the machine roam around the room and program itself to respond to events that upset its view of the environment, I began realizing that yet another key element was missing from the theory. It wasn't until I simulated the intelligence on the screen of a home computer that I found that missing element.

Rodney does something that is difficult to notice whenever you don't know exactly what to look for. Rodney exhibits a form of adaptive intelligence that is more clearly observable by means of a faster-than-life simulation. The creatures have the ability to adapt their behavior patterns to meet changing needs. It was there all along, but I missed it.

So the Rodney machine was still new and the manuscript for *How To Build Your Own Self-Programming Robot* was still in the editorial stages of publication when I began work on this book. Now the perspective is clear, and I do not have the feeling that anything is missing.

This book on computer simulation of robot intelligence does not make the Rodney project obsolete, however. As mentioned earlier, Rodney exhibits the qualities of adaptive machine intelligence. The advantage of using the computer simulations is that it is far easier to study the mechanisms, gather data, and compile graphs that show the inherent nature of evolutionary adaptive machine intelligence.

A secondary advantage of this simulation project is that one can begin doing useful research work without having to spend a number of months building the hardware from scratch. Of course, a little spot of light blinking around on the monitor of a home computer lacks the flavor of a mechanical gadget whizzing around the room. However, robot experiment who have no taste for building hardware can participate in the work that has to be done, putting their home computers to work.

The past six years have amounted to a search for meaningful generalities. The objective has been to come up with a set of theories and perspectives on robotics that transcend the current state of electronics technology and thus promise to be useful for a good many years to come. For the first time, I have the distinct impression that we have arrived.

Future developments in electronics and computer technology will certainly make the methods for demonstrating and implementing the theories different from what they are now. The work is not contingent upon future developments; everything is here now.

This is not to say that the bulk of the work is already done. Far from it. Every detail must be checked, doublechecked, and backed up with endless hours of careful thought and hard work. While there might appear to be few, if any, serious loose ends on the basic theory, a lot of questions remain to be answered, and the means for answering those questions (or indeed many of the questions, themselves) cannot be framed until some of the preliminary work is done.

THE THEORY OF EVOLUTIONARY MACHINE INTELLIGENCE

In a nutshell, the whole idea of evolutionary machine intelligence is to develop machines in the context of machines and leave the theories of electrical activity in biological tissues to the physiologists. Without a doubt, computer technology is enjoying a level of power and sophistication beyond that achieved by any other manmade device. It is instructive to imagine what the state of computer technology would be today if everyone had insisted upon developing computers along the theories of the human brain. We would probably still be trying to culture an AND gate in a test tube. So the notion of letting machines develop in their most natural context is not a new one at all. For reasons that seem to surpass any rational human understanding, however, roboticists have traditionally failed to view things that way.

Looking at electronic circuits in their most elementary forms, you will find that many of them respond to relevant influences in a reflexive fashion. A loudspeaker, for instance, responds to signals from a microphone, a light responds to the action of a switch. The scheme has an input and an output. Do something at the input that is relevant to the mechanism and something happens at the output.

Fitting the idea into the framework of very simple robot machines doesn't create any technical or philosophical problems. Quite often it is a simple matter of changing the wording. A robot can sense contact with an obstacle and respond by moving away from it. In a very real sense, this is an example of machine reflex activity. Why get all tangled up with the theories of nerve conduction and muscle activity when a 50-cent switch, a couple of digital ICs, and a motor can do the same job?

Reflex activity, in a purely electronic context, can be carried to some very elaborate extremes. A great number of simple sensing devices, coupled through a logic control scheme, can feed vast amounts of relevant information to a great variety of response mechanisms. The whole process can be purely reflexive, employing little logic other than that necessary for controlling the flow of data.

In the framework of evolutionary machine intelligence, a purely reflexive mechanism represents the simplest sort of machine. For the sake of keeping things straight, I have assigned neat-sounding Greek letters to the various stages of robot evolution. So a machine that is purely reflexive in nature is an Alpha-Class machine.

My Buster robot was an *Alpha-Class* robot. Its activities were restricted to dealing with conditions in the immediate environment. The creature existed in the moment and at its particular place in the environment. Any notions of past or future experiences or environmental conditions other than those of the moment, are wholly irrelevant.

It is very tempting, at this point, to draw an analogy, between a simple Alpha-Class machine and the reflex arc found in all vertebral animals. But remember, that drawing such analogies at this level is bound to get you into trouble. The trouble here is that the vertebral reflex arc exhibits programmed qualities. Accidentally touch your finger on the heating elements of a hot stove and your reflex arc will invariably call for a reflex action—jerking your finger away.

This sort of programmed activity is far too restrictive in the context of robot machines. The moment you begin telling the machine exactly what it is supposed to do, how it is suppose to respond, you are asking for a whole lot of trouble. Consider the robot experimenter crying for programming help, simply because he finds the prospect of programming his robot to repond to all possible contigencies a task of staggering proportions.

It doesn't take a whole lot of imagination to see that the idea of programming a robot to deal with all possible environmental contingencies soon runs headlong into monsterous difficulties. It's precisely difficulties of this nature that signal a need for approaching the situation from a fresh viewpoint. The fresh viewpoint in this case is so simple that we have to pinch ourselves to see whether or not it amounts to a nice dream—let Alpha-Class machines respond randomly.

All you must do is tell the machine it is supposed to do *something* whenever a relevant change occurs in the environment, then let the machine figure out, at random, what sort of response allows it to adapt to that change.

To be more specific, suppose you build a robot machine that is capable of sensing an obstruction in its path and responding by moving in any of 16 different directions. It first senses the obstacle, and being programmed to respond at random, begins trying various motions. Some will work and some will not. The first one that works clears the problem and the machine is free to roam around again.

So what if the machine has to try maybe four different responses before it discovers one that works? The result is the same as a different sort of machine you programmed in advance to make the right move the first time. Maybe the random-reflex creature wasted some time and effort, but it does the job and without using any memory other than that required for sensing that something has to be done!

Thus, an *Alpha-Class machine is one that makes purely random and reflexive responses to conditions in its immediate environment.* It is a very real part of evolutionary intelligence because it represents the first, elemental step in a heirarchy of machines. It fits the adaptive nature of machine intelligence because it is capable of generating new responses to meet new conditions.

The complexity of an Alpha-Class machine can be multiplied to any desired extent. There can be any number of sensing mechanisms and any number of response modes at its disposal.

As long as the responses are purely reflexive and random in nature, it remains an Alpha-Class machine.

The notion of building an Alpha-Class machine with a vast variety of sensory inputs and response mechanisms brings up an intriguing question. Is it possible for an Alpha-Class machine outfitted with a variety of high-resolution sensory inputs and equally elaborate responsive outputs to appear to behave in a quasi-rational fashion? The question must be answered one of these days, and perhaps you are the one to do it, not in words, but in deeds.

Also, for the sake of clearer communication between robot experimenters, I have devised a scheme for classifying Alpha-Class robots even further. Using Roman numerals as a guide, an Alpha I machine is an Alpha-Class machine that has only one sensory mode and one response mode. An Alpha II machine on the other hand, has either more than one sensory mode, more than one response mode, or more than one of both.

An example of an Alpha-I machine is one that senses only obstacles in its path and responds by finding a way to get around them. An Alpha II, however, might sense both physical obstacles and bright lights, responding in either case with some motion activity.

In this book, you will find computer programs for building both Level-I and Level-II Alpha-Class creatures.

A *Beta-Class* machine is simply an Alpha-Class mechanism equipped with a memory of past experiences. The machine exhibits Alpha-like behavior at first, but each time it stumbles across a response that works it remembers that response, filing it away at an address location determined by the nature of the condition that elicited the response. So when the Beta creature encounters the very same condition again in the future, it looks to its previous response and tries it again.

The overall impression is that the Beta-Class machine behaves in a random fashion for a while. With experience, however, the random, Alpha-like behavior begins giving way to predominately Beta-Class behavior. The creature makes fewer and fewer calls upon its random response mechanism, interacting with its environment in a more efficient manner.

Adding a memory scheme to an Alpha-Class mechanism, thereby creating a new order of machine, is a very natural and logical step in the development of machine intelligence. It is difficult to imagine any other sort of step that would be simpler and yet exert such a change in the creature's mode of behavior.

A Beta-Class machine is preprogrammed to do only two things: to respond to relevant changes in its environment and save workable responses in a memory. Exactly how the machine responds and exactly what gets stored in the memory is largely beyond the experimenter's control or, for that matter, the experimenter's concern.

If the Beta machine is further equipped to change its mind as exact circumstances change, it is fully prepared to deal with the environment in an adaptive fashion. The Alpha mechanism is always there in the background, resting below the memory and coming to the rescue whenever the creature encounters a new situation or finds that an old response no longer works.

The difference between a Beta-I and Beta-II machine has to do with the number of input and output modes. Beta Is use a single sensory mode and a single response mode of expression. A Beta II has more than one of either sensory or response modes, or more than one of each.

A Level-III Beta creature, incidentally, would be one that has more than a single-Alpha-Class function included within it. A Level-III robot creature, for example, might have a single processor devoted to Beta-like learning activity, but several more Alpha schemes for controlling manipulators or other external gadgets.

Other than an evolutionary and adaptive perspective, there is nothing really new about these Alpha and Beta machines. The history of Alpha-Class machines goes all the way back to the days when physiologists were constructing electronic models of simple biological creatures, and the Beta-Class history reaches back to the 1950s when researchers and high school science students were building the first mechanical mice for solving mazes.

So in the evolutionary scheme of things, what comes after adding some memory to a system that is capable of responding to changes in its environment? Look at it this way. A purely Alpha-Class machine exists in the moment. It has no way to work with events of the past or future. A Beta-Class machine can call upon successful solutions to past problems in order to deal with the problems of the moment more effectively. What's missing? What is missing is teh creature's ability to anticipate events that might occur in the future.

A *Gamma-Class* machine is one capable of generalizing what it knows from first-hand experience to similar conditions not yet encountered in the environment. The machine works out sets of "educated guesses" regarding the nature of possible situations in its future, studies its own past experiences, and generalizes relevant elements of those experiences, saving them for a time when they might be needed.

The Gamma generalization scheme is built right on top of a Beta mechanism which, of course, is built upon the primitive Alpha mechanism. So even a Gamma-Class robot begins its life in an Alpha fashion, blundering its way around the environment, remembering those responses that work. And like a Beta-Class machine, Gamma soon behaves in a more direct manner, its Alpha random behavior slipping away in favor of learned responses.

However, something else goes on in the "mind" of a Gamma-Class creature. While it is in the process of building up confidence in its learned, sucessful first-hand experiences, it is conjecturing about ways to deal with other kinds of events. Instead of resorting to purely Alpha-Class behavior when it encounters a brand-new situation in the environment, a Gamma-Class creature has a set of responses already residing in memory.

Those Gamma-generated responses are bits and pieces of knowledge gained at various stages of its earlier life. The conjectured response is immediately tried whenever the occasion arises. If this response happens to work, the creature's confidence grows and it has additional information for firming up its notions about dealing with other future events. But in the event a conjectured response is partly or, in some instances, altogether wrong, the creature resorts to Alpha reflex activity.

The memory of a Gamma-Class robot is in a state of continuous fluctuation and change, at least as long as it is interacting with a rich and dynamic environment. Put the machine into a sterile environment and you will find very little Gamma activity taking place. As a result, the creature will be largely unprepared to deal with unforseen circumstances.

So the evolutionary heirarchy begins with the primitive Alpha-Class machine and builds quite logically to the Beta- and Gamma-Class machines. The evolutionary procedure suits the context of electronics quite well, and there is no need to make any conscious references to basic biological activity. If, after you've had a chance to investigate these mechanisms for yourself, you find the whole idea has some staggering implications, consider this—it is all done in a very simple fashion. If there is anything more intriguing than the system itself, it is its simplicity.

I have never calculated the amount of programming required for various tasks demonstrated in this book, but I'm willing to bet that at least half of it would be classified as housekeeping chores for your computer. These would include the steps necessary for printing out various kinds of data on the screen, drawing the images and getting test information in and out of the system.

Another major portion of the programming has to do with giving the creatures some sensory and responsive mechanisms. *There is absolutely no programming devoted to telling the creature exactly what it is supposed to do under any particular set of circumstances.* An estimate of 25 man-years to program a machine to respond to all possible contingencies is time and effort down the drain. Let the machine program itself as the need arises, and spend all that time dreaming up new ways to extend the machine further and test the capabilities it has. Simplicity is the key to success in any technological and scientific endeavor.

EAMI AND ANIMAL BEHAVIOR

Through this chapter, I have made repeated and largely negative references to the notion of drawing analogies between biological activity and the fundamental workings of robots. It is not my intention to demean work that has been carried out with clear-cut objectives and scientific responsibility.

Rather, my intention is to drive home the point that the EAMI concept relies entirely on the notion that intelligent machines must evolve in the context of machines, and not that of physiology. If I have made that point clear to you, and you now understand that even casual references to biological analogues can, at this point, cause more confusion than anything else, we can all forget about knocking the efforts of physiologically oriented researchers.

In fact, the EAMI concept complements the endeavors of other kinds of robot research, once you get past the fundamental levels, that is. One day soon, EAMI will develop to the point where we are going to seek out the analogue builders with hat in hand. We are going to need some of their elaborate sensory and responsive mechanisms to transform the fundamental concepts of EAMI into working machines.

Of course, EAMI re-searchers will not apply the physiological models the same way, and there will be a need for simplifying many of these borrowed mechanisms. But as long as there are researchers who insist upon doing everything from a biological viewpoint, we ought to get along with them.

So an EAMI machine must grow up in a purely machine-oriented context. But even so, there comes a point where it is tempting, and even desirable, to begin drawing parallels with the animal world. Now I did not say we will be building EAMI machines that mimic animal behavior, I said "drawing parallels."

You won't have to work your way very far into this book before you begin noticing forms of behavior that appear quite animal-like. It is at the behavioral level that parallels beween EAMI and the animal world become legitimate.

If we are to seek ideas outside the realm of electronics and computer science, the appropriate place to look is the lab of the animal psychologist. Indeed, if you have any acquaintance at all with the fundamentals of animal psychology, you will find some familiar expressions scattered through the experiments in this book. You will find terms such as *learning curve, adaptive behavior*, and *behavioral reinforcement*. In fact, the use of the word *memory* in this book refers more often to the psychologist's view than to the engineer's.

EAMI is inseparable from the science of psychology, simply because it is a behavior mechanism applied to machines. In a sense, we are far more interested in dealing with machine behavior and psychology than the mechanisms required for doing the job. It just so happens that EAMI is the simplest, most efficient, and effective way to get a machine operating at a level where its psychology can be studied in a meaningful way.

Bear in mind, however, that we are not building psychological models either. Building a creature that does nothing but solve mazes defeats the purpose of the whole effort and puts us right back into the seat of robotics 25 years ago.

Like the animal psychologist, we have a creature sitting in front of us that demands study in its own right. The creature already exists, both with its strengths and its shortcomings. Our job is not to build a machine that does a particular set of tasks; that's old hat. The idea is to study a creature that already exists and learn as much as possible about its behavior. The work is more fun that way and a whole lot more meaningful in the long run.

A WORD ABOUT DEMONSTRATING EAMI SYSTEMS

We all do things such as build robots or simulate robot intelligence on a computer for various personal reasons. One such reason, I think most of us have to confess, is that of impressing other people with the things are able to do. For your own sake, I hope that is not your most compelling reason for wanting to simulate robot intelligence.

The problem with trying to impress other people with this idea is that it doesn't impress them at all. The whole affair comes across looking like some sort of video game, and a cheap one at that. If your enthusiasm for the project is founded upon the idea of impressing other people, you are in for a terrible disappointment.

The only people who will truly appreciate what you are doing are those who understand the problem you are solving. Most folks want to see gadgets clanking around the room, responding to voice commands and doing neat, but banal, things such as washing the dishes.Show them a little blinking square of light bouncing around on a computer screen, exhibiting the germ of innate machine intelligence that marks a bridge between living and non-living things, and they'll wonder whether you have lost your marbles.

The situation has gotten so bad in my own case that I no longer demonstrate my robots and intelligence simulators unless the observers are willing to sit still for a crash course in EAMI, something along the lines of the material presented in this chapter. Even then, people walk away, wishing they had seen a thing sweeping up the mess of papers and bits of electronic parts that generally clutter the floor in my office.

I suppose we are facing the same sort of situation that plagued film producers in the very early days of cinema. Many people seeing a movie for the first time in their lives thought the whole thing was a hoax. It sounds naive to us these days, but there was a prevailing impression that the moving images were being created by a whole cast of players working behind that big, flat, white sheet on the stage.

There is great potential for hoaxes in the robot business these days, and there are some pretty smart people who are going to give the public exactly what they want, snazzy machines that add up to nothing. And, that's going to make things even tougher for those of us who are working diligently along the proper path.

I suppose the thought is potentially depressing, but it certainly does not make serious robot work a waste of time and effort. We know what we are doing, and that's really what counts. Besides, it's a lot of fun and it offers some very thrilling moments. That counts for something, doesn't it?

Getting The Most From
This Book And Your Work

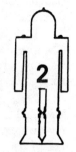

The primary purpose of this book is to give you a chance to see artificial intelligence at work on the monitor screen of your personal computer. It isn't all a matter of simple demonstration, however. You will have plenty of invitations to play some games with the creatures, invent new games, and do some scientific research and experiments.

I suspect the people reading this book run the gamut from very beginners at programming computers to both amateur experts and professionals. As you might suspect, it is quite a challenge to prepare a book that is both meaningful and interesting to such a broad spectrum of users.

Then, there is the challenge of writing a book—one that is essentially a book about using computer programs — for people who might not have access to the same computer I happen to use. Well, here's how it's done.

LAYOUT OF THE PROJECTS AND PROGRAMS

This book can be divided roughly into three parts. The first part deals with Alpha-Class creatures, the second with Beta-Class versions, and the third part deals with Gamma-Class creatures. That's pretty simple.

After you see that the book is divided up by creature class, you'll then find Level-I and Level-II versions. This classification has nothing to do with Level-I and Level-II BASIC for the TRS-80 personal computers. All of the work assumes your computer has the capability of Radio Shack's Level-II BASIC. What these expressions mean is Alpha I and Alpha II, for example. Sorry about the potential for confusion.

So the book begins with work on Alpha I creatures and then moves to Alpha II versions. Then comes some Beta I projects, followed by a few Beta II projects. Finally, you will find some

Gamma I projects and a few hints on programming a Gamma II creature.

That is the general scheme of the book. It begins with primitive Alpha creatures and moves gradually through Beta and Gamma versions. In other words, it follows the same evolutionary process outline in Chapter 1.

Leafing casually through the pages of this book, you will see a whole lot of computer program listings. I have discovered from my own experience that entering programs into a computer can get to be a very boring and tedious task, when it all has to be done by hand at the keyboard, that is.

In order to relieve some of this programming tedium, the programs have been designed so that they can be entered in smaller, bite-sized pieces. You will find an extensive use of subroutines and their corresponding GOSUB statements. Organizing the programs around so many subroutines makes it possible to work out new programs based on large sections of earlier programs. The result is that you rarely have to enter an entire program from scratch. Here is how it works.

Suppose you are getting started in Chapter 3, the first chapter that contains any Alpha-I programming. This doesn't happen to be a very extensive program, and being fresh on the job, you probably won't be too frustrated by having to enter it from the keyboard in its entirety.

After you get this first Alpha-I program into your machine, tested and debugged (more to say about testing and debugging in a moment), you are ready to play with it. Here's the important point—before going to the next program in the sequence, save the first one on cassette tape, then load the new program "on top of" the older one, not on cassette tape, but in the computer memory.

All but about four of the programs in this book are built around portions of other programs which can already be residing in program memory when you begin entering them. Maybe 5 or 6 subroutines called by one program can be used in the next program. There is no need to enter them all from scratch. Simply build the new program around the subroutines already in place.

The instructions for entering a new program clearly specify which one of the older programs must be entered from cassette tape first and then specifies which of the older subroutines are to be used and which ones should be deleted. What could be a terribly long and tedious procedure thus becomes a simple set of programming steps.

Using this idea of duplicating subroutines, building the bulk of one new program around subroutines for an older program forces you to do a couple of things, however. First, you cannot expect to take advantage of this time-saving technique if you skip around in the book. You must enter the programs as they are presented here. What good is the technique if you skip to a program that calls for building around subroutines from an earlier program you haven't done?

Perhaps the other demand is obvious, you must save each completed program on cassette tape. Not all programs are built around the previous one. Some call for loading programs found a couple of chapters earlier in the book. So you ought to be prepared to save every program on cassette tape and keep an accurate index of where those programs are on the tape. In other words, work the programs in the exact order they appear in this book and save all of them on cassette tape.

Now for this matter of testing and debugging the programs. Some of the programs are rather short and simple, but of course this isn't always the case. Entering so many programs brings up the distinct possibility of making errors and many kinds of errors will "bomb" the whole program.

For the sake of novices who haven't done a whole lot of computer programming, you will find some specific steps for testing programs and subroutines. Whether or not you think you have entered the program with absolutely no errors, it is a good idea to follow the suggested testing routines. That way, you will be able to uncover errors before they become lost in a large listing and cause real problems.

Since all of these programs have been carefully tested and proven, the general procedure for working out an error is comparing your program listing with the printouts appearing at the end of each chapter. I must say, however, that I cannot take the responsibility for errors if you fail to follow the instructions to the letter.

There is more to this business of doing research with machine intelligence than entering the programs and watching the critters do their thing on the screen. I am convinced that you must also understand exactly how the programs work, right down to the *tiniest* detail. If you do not take the trouble to understand how the programs work, you cannot possibly make a convincing case for anything we have to say about evolutionary adaptive machine intelligence.

To this end, most of the text in each chapter is devoted to the theory of operation of the programs. The programs are first presented in the form of flowcharts, followed by segments of programs and subroutines that are directly related to the blocks on the flowchart. You thus have a chance to study the program in bite-sized pieces, relating it to the flowchart and hopefully understanding what each line of the program does and why it does it.

So when you begin studying a new program, you will find a flowchart, some theory of operation, and then a complete listing of program lines that are either new or unique to that particular program. Subroutines that are borrowed from earlier programs are listed by subroutine numbers and names, but neither listed nor discussed again in any great detail.

Having the programs broken up into bits and pieces makes it much easier to study and enter into your computer. At the same time, though, the procedure makes it difficult to doublecheck a complete program listing. For this reason, you will find complete program listings at the end of each chapter. Any problems, errors, and points of confusion, which might be hidden away in the main text of a chapter, can be readily resolved by comparing your entire program listing with the version printed at the end of the chapter.

I am sure it will occur to some experimenters that they can enter the programs from the composite listings at the end of the chapters. Please don't do things that way. Not only will you end up doing a lot more work than necessary, but also run the risk of playing around with programs you don't understand. The whole purpose of the composite listings at the end of the chapters is to diagnose any errors that might crop up.

AN INVITATION TO EXPERIMENT AND PLAY

Much of the fun and excitement of this whole subject of machine intelligence would be lost if all the work is presented as straightforward demonstrations. Demonstrations, as such, might have some educational value, but they rarely stir a sense of lasting excitement. The work in this book is designed to let you have some fun with the creatures and look forward to doing some original research of your own. Then too, it is possible to write programs that get the creatures to do some fascinating and sometimes humorous sorts of things. While there is a certain brand of fun and excitement associated with doing serious research, I see nothing wrong with pursuing some fanciful ideas for the fun of it.

Scientific research should never be the sole property of one individual. Being in the forefront of this matter of evolutionary adaptive intelligence, I have the chance to do a lot of research that has never been done before. This does not mean it is a waste of your time to duplicate the work, however. In fact, I encourage every experimenter to duplicate the experiments, refine my conclusions, and look for loopholes that might have escaped my attention. I also encourage you to extend the work as far as you can. I'm not going to get all uptight about anyone publishing articles and writing books about projects I could have done myself. You shouldn't worry about others "beating you to the punch," either. There is far too much work to be done to mess around with trivial matters such as jealousy. All I ask is that you carry out original research in the same spirit I present it here — carefully, thoroughly, and with a sense of responsibility.

In the final chapter of this book, I outline rather clearly the things I plan to do next with the EAMI concept. I invite you to follow the same path, and if you happen to get the job done before I do, that's all right. We can compare results, work out any discrepancies and go on from there.

In that final chapter, I also suggest some avenues that have to be opened sooner or later, but ones I do not intend to pursue myself. Get moving gang! Work your way through the theories, projects, and programs spelled out in this book, and then start doing your own thing. Let's see if we can help make robotics something beyond a tinkerer's art by 1985.

THE NEED TO CONSERVE MEMORY SPACE

At first glance, many experimenters might find the BASIC programs looking a bit strange. The program lines are written without the benefit of any spaces between words, expressions, and characters; and what's more, most of the lines are multiple-statement lines.

The reason for doing this is to save valuable program memory space that would otherwise have to be devoted to characters for spaces and line numbers. Getting the programs to fit into the smallest personal computer system is more important than writing programs in a way that simply makes them a bit easier to read.

It doesn't take a lot of experience with these space-saving techniques to see what is going on, however. If you find yourself getting confused by the lack of spaces between some of the words and characters, there's no harm in dividing them with a light pencil stroke in the book. You will get accustomed to reading the crammed-together program formats after a while.

A WORD TO READERS WHO DO NOT HAVE A TRS-80

All of the programs presented in this book assume you are using a Radio Shack TRS-80 personal computer with Level-II BASIC and at least 4K of memory. The choice of this particular machine is based on cold facts. No matter what your opinion of that machine might be, it is a matter of record that there are twice as many TRS-80 users as there are Apple and PET users combined. If you happen to be using one of the more sophisticated S-100 oriented systems, you probably know enough about computer programming to make the necessary adjustments anyway.

Apple and PET users should not feel completely left out of this project, however. While these machines employ the BASIC language, there are some small but significant differences. To this end, every program is presented in the form of a flowchart. The flowcharts are portable; that is to say, they apply to any kind of machine that uses any sort of BASIC.

So if you aren't using a TRS-80, you can use the flowcharts as guides for devising the appropriate programs. One of my own immediate objectives is to produce an Apple version of this book, but don't wait around for me to do the job for you. For one thing, there is a terribly long delay from the time one makes such a program and the time it appears in print. Then too, you will miss out on the chance to learn a lot about making translations between the languages of two different machines, a talent that is badly needed in both amateur and professional circles these days.

Alpha-I Basic — The Starting Point

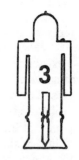

3

Just as the entire scheme of adaptive machine intelligence is an evolutionary one, so is the process of building up simulation programs for demonstrating it. It all starts with the ALPHA-I BASIC described in this chapter and then it builds up from there. Every program described in this book has some ALPHA-I BASIC features in it. So the work you are about to do now has relevance to everything you'll be doing in the future.

Since ALPHA-I BASIC is the foundation to all other work, it is most important to understand how it works and what it means. If you are careless with your study and programming at this point, I'm afraid you will have to pay for it dearly later on. Study the overall concept, the subroutines, and even the individual lines and statements carefully. They will all be used time and again in later programs, but without the benefit of extensive explanations.

THE NATURE OF ALPHA-I BASIC

ALPHA-I BASIC is the simplest possible form of adaptive intelligence. The creature's purpose is to move about in a well-defined environment, doing whatever it can to ensure freedom of motion within that environment. Its responses are purely random in nature with only two restrictions upon its behavior. First, the creature must remain within the confines of its environment, and second, it must not be allowed to stop moving.

Both of these restrictions have their counterparts in the real world. The real world, for instance, does have barriers of some sort that limit the range of a creature's activities. These might be physical barriers such as walls, the surface tension of a drop of water, the forces of gravity that confines creatures to the surface of a planet, or, in the most esoteric sense, the abstract confines of a 3-dimensional space and the forward-pointing arrow of time.

The creature represented by ALPHA-I BASIC is confined to a rectangular field drawn on the screen. All the creature ever perceives and does is restricted to the confines of this rectangle — existence outside this world rectangle has no relevance whatsoever. The creature's activities must be limited to the area enclosed by the rectangle. And, in the case of ALPHA-I BASIC, its total mode of behavior is one of avoiding the possibility of leaving that little world.

In fact, the purpose of the creature in ALPHA-I BASIC is to actively engage the barriers of its little world, and the need for actively engaging the barriers precludes the possibility of standing totally motionless. Motion is the essence of the creature's behavior and existence, and allowing it to stand motionless within its world defeats its purpose for being created in the first place.

As far as the ALPHA-I BASIC creature is concerned, the need to pursue continuous motion is built into its very nature. This will not be the case for some ALPHA-II and some of the higher-order creatures (standing motionless can, indeed, be a legitimate and useful response where the range of perceptions and modes of behavior are sufficiently complex).

In a few words then, an ALPHA-I BASIC creature is characterized by continuous and random motion within the confines of a rectangle. Its purpose in "life" is to pursue freedom of motion, and the adaptive quality of its behavior is one of finding ways to keep moving in spite of changes in its barriers.

THE ALPHA-I BASIC SCREEN FORMAT

Figure 3-1 shows the screen format for the ALPHA-I BASIC experiments. It is simply a rectangle drawn on the screen and a small square representing the creature, itself. Incidentally, the drawing is shown here with blacks and whites reversed. The border and creature figures are both white against a black background. For the sake of showing clearer pictures, this convention is used throughout the entire book.

THE ALPHA-I BASIC FLOWCHART

The general flowchart for ALPHA-I BASIC is shown in Fig. 3-2. It is a rather simple flowchart, but since it will be incorporated in virtually every experiment described in this book, it is one worthy of some very close study at this time.

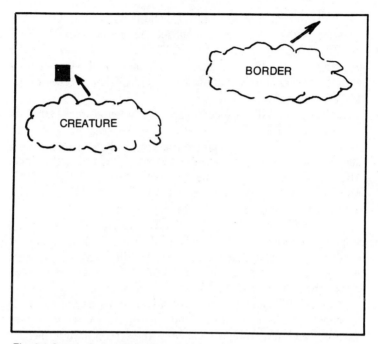

Fig. 3-1. Screen format for ALPHA-I BASIC.

The program begins with a short heading that, for the sake of the experimenter, describes the nature of the program. In this case, the heading does little more than identify the program as one for ALPHA-I BASIC.

Upon displaying the heading, the experimenter initiates the active part of the program by satisfying the START conditional. The heading remains displayed on the screen as long as the "answer" to the START conditional is N (*no*). In this particular case, striking the ENTER or RETURN key satisfies the START conditional and causes the DRAW FIELD operation to take place. DRAW FIELD is responsible for drawing the rectangle on the screen. It's nothing more than that.

The INITIALIZE step merely gets the creature started in its little life on the screen. As you will see later, INITIALIZE begins the creature's motion near the upper left-hand corner of its rectangle world and causes it to move downward and slightly to the right. Actually, INITIALIZE, simply sets up the creature for its first incremental move across the rectangle. SET NEXT MOVE compiles the initial information and translates it into a specific motion code.

Next comes the CONTACT conditional. The CONTACT question is this: Will the next move, the one compiled by SET NEXT MOVE, make me run into a barrier? The answer (Y or N) determines what happens after that. If there will be no contact with the border, the N route from CONTACT jumps down to the MOVE operation.

MOVE actually executes the incremental move that is compiled by SET NEXT MOVE. Since the creature is free to move, the little white square representing the creature moves one incremental step across the screen. After doing this little MOVE, the system sets up the next move at SET NEXT MOVE, the move is checked for CONTACT, and if there will be no contact, the creature moves yet another step.

The SET NEXT MOVE, CONTACT, and MOVE looping sequence continues until CONTACT shows a Y condition, a condition indicating that the next move will result in a contact with the border figure. The creature, in other words, moves across the screen in a regular, stepwise fashion until it finds the next move will cause a contact situation. As far as the observer is concerned, the creature square seems to move rather smoothly and in a straight line across the screen until it encounters the edge of its world.

The speed and direction of motion is mainly determined by SET NEXT MOVE. When the program is first started, the parameters for SET NEXT MOVE are dictated by INITIALIZE. As mentioned earlier in this section, INITIALIZE sets the first element of motion from the upper left-hand corner of the border and down toward the middle of the lower boundary. This initial motion is the only one that is dictated by the program. All motion thereafter is determined by the machine itself.

So far in this explanation, the creature has been initialized and moved in a stepwise fashion toward the bottom of the border figure. The SET NEXT MOVE, CONTACT=N, and MOVE sequence remains in force until the creature makes its first contact with the border figure. Now suppose this first contact has indeed been anticipated for the next move. The answer to the first CONTACT conditional is Y, and the system enters a different operating loop composed of BLINK, FETCH NEW MOVE, DECODE, and a second CONTACT conditional.

So upon seeing the next move would result in an undesirable contact with the border figure, the flowchart shows a BLINK — the creature figure is simply blinked on and off one time.

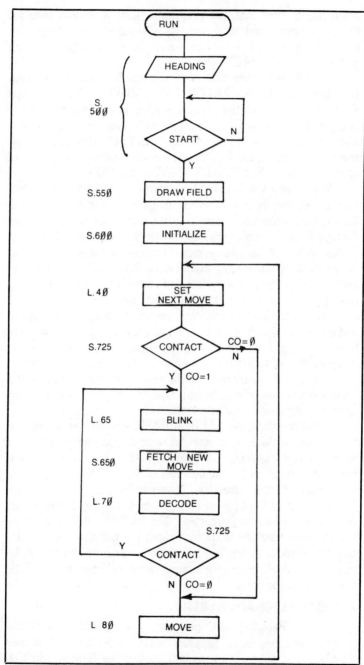

Fig. 3-2. General flowchart for ALPHA-I BASIC.

Immediately after that, the program selects a random code for the next speed and direction. As described a bit later, the randomly generated motion code must be decoded before it can be tested for CONTACT again.

If the new motion code results in another contact, the loop is repeated: BLINK, FETCH NEW MOVE, DECODE, and CONTACT. This motion-code selection loop remains in force until it stumbles across a motion code that will not cause a contact. The answer to the second CONTACT conditional is N, and the system enters its original motion-producing sequence of operations: MOVE, SET NEXT MOVE, and CONTACT.

When the experimenter first starts the program, the creature figure can be seen moving diagonally across the screen. This apparent motion continues until the creature seems to make contact with the lower part of the border. At that moment, it blinks on and off an undetermined number of times, maybe just once or maybe six or even eight times. It just sits there blinking each time it picks a new motion code until it happens to pick one that carries it away from the border figure.

With selection of a randomly generated motion code (speed and direction) that frees it from the border, the creature appears to move along until it runs into the border again. At this point, there is no way to say exactly where the creature will be going. After the initial excursion from the upper left-hand corner of the border figure, there's no telling how the creature will respond upon sensing contact with the border.

While watching the ALPHA-I BASIC creature in action, you might want to think of an amoeba scurrying around in a drop of water. It is a vigorous and mindless activity which is confined, and largely dictated by, the confines of the surface tension of the water drop. At the same time, however, be sure you do not fall into the trap of thinking you have something that is mimicking the behavior of a living organism. The ALPHA-I BASIC creature is a machine creature in its own right; the fact that its behavior appears similar to that of an organic creature must not be allowed to confuse the real importance of what you will be observing on the screen.

SUBROUTINES FOR ALPHA-I BASIC

The ALPHA-I BASIC program is composed of five subroutines and one main program which pulls the subroutines together. The subroutines, as indicated on the flowchart in Fig. 3-2, are as follows:

S. 5ØØ HEADING/START
S. 6ØØ INITIALIZE
S. 725 CONTACT
S. 65Ø FETCH NEW MOVE

Of course, the main program will call these subroutines at the appropriate times by executing a GOSUB *line number* statement, where *line number* is the designated subroutine number—5ØØ, 6ØØ, 725, and so on.

The flowcharts for later program will use many of the same labels for the operational blocks, and, in fact, the same subroutine line numbers. This procedure allows you to relate blocks on the flowchart directly to program subroutines. As a matter of further convenience, an S. 6ØØ INITIALIZE block appearing on one flowchart does essentially the same job as an S. 6ØØ INITIALIZE block does on some other flowchart in the book. So once you understand the main purpose of a subroutine, you can generalize the notion to other programs that use it.

However, the exact program statements contained in a subroutine often varies somewhat from one program situation to another. Although two different programs might have flowcharts showing S. 6ØØ INITIALIZE, and although they perform essentially the same task, the programming might be slightly different. For this reason, the subroutines actually entered into the computer carry slightly different names.

For example, the S. 6ØØ INITIALIZE block in this flowchart carries the REM name INITIAL-1. A flowchart in the next chapter also shows a block labeled S. 6ØØ INITIALIZE, but since the actual programming is slightly different, its REM name is INITIAL-2.

While this nomenclature might seem to be rather awkward or confusing at this time, it is only because it is an attempt to deal with the whole project at once. The idea is to simplify matters by using the same flowchart designations and GOSUB line numbers for similar kinds of operations throughout the book. If you do not grasp the significance of this right now, you will find yourself taking advantage of it anyway. Sooner or later you'll catch on.

Subroutine 5ØØ HEADING/START

This is the heading subroutine for ALPHA-I BASIC. To distinguish it from other 5ØØ HEADING operations, it is designated HEADING-1. In line 5Ø5, HEADING-1 clears the screen.

After that, it displays the messages RODNEY ALPHA-I and BASIC VERSION (lines 51Ø and 515).

The message, DO ENTER TO START, is part of an INPUT statement in line 525, and the only way to get out of this subroutine is by striking the ENTER key. Doing so, the value of string variable S$ is defined as the null string, but since it isn't used anywhere else in the program, the actual definition isn't relevant. Line 525 merely gives the experimenter a chance to rest in the HEADING/START condition until striking the ENTER key.

```
500 REM ** HEADING-1 **
505 CLS
510 PRINTTAB (25) "RODNEY ALPHA-I":PRINT
515 PRINTTAB (25) "BASIC VERSION"
520 FORL=ØTO1Ø:PRINT:NEXT
525 INPUT"DO ENTER TO START";S$:RETURN
```

Variables introduced: L, S$

Variables required at start: none

Test Routine
```
10 CLS:GOSUB 5ØØ
20 CLS:PRINT"OK"
30 INPUTX$:GOTO1Ø
```

As indicated here, HEADING-1 introduces variables L and S$. Since HEADING-1 is executed only one time, when the program is first started, these two variables can be redefined and applied to other functions anywhere in the program. HEADING-1 requires no variables at the start. It can be run, in other words, without the need for defining any variables.

The test routine for HEADING-1 is a rather simple one. After entering HEADING-1 into your system, using the line numbers designated here, it can be tested with the specified test routine. Enter the test routine and then enter the RUN command.

If all has gone well, you should see the heading messages printed out neatly on the screen. Striking the ENTER key causes the program to return to line 2Ø of test routine which clears the messages from the screen and prints OK in the upper left-hand corner.

According to line 3Ø of the test routine, the INPUTX$ allows you to repeat the test sequence any number of times by simply striking the ENTER key again.

If you're ready to go to work on ALPHA-1 BASIC, get HEADING-1 into your machine and check it out. Be sure to delete the test routine when you're satisfied the subroutine is working properly.

Subroutine 550 DRAW FIELD

The purpose of S. 550 DRAW FIELD, wherever it might appear in this book, is to draw the creature's rectangular border. The configuration of the border varies somewhat in later programs; so although S. 550 DRAW FIELD appears on virtually all the main flowcharts, the actual subroutine program has to carry unique names. In this instance it is called FIELD-1.

```
550 REM * FIELD-1 **
552 F1=15553:F2=15615:F3=16321:F4=16383
555 FORF=F1TOF2:POKEF, 131:NEXT
560 FORF=F3TOF4:POKEF, 176:NEXT
565 FORF=F1TOF4STEP64:POKEF,191:NEXT
570 FORF=F2TOF4STEP64:POKEF,191:NEXT
557 RETURN
```

Variables introduced: F1, F2, F3, F4, F
Variables required at start: none

Test routine
```
10 CLS:GOSUB550
20 INPUTX$:GOTO10
```

This drawing subroutine uses TRS-80 POKE graphic to draw the border configuration shown in Fig. 3-1. Line 552 defines the size of the rectangle, while lines 555 through 570 actually draw it on the screen. In this case, line 555 draws the top of the border, line 560 draws the bottom, and lines 565 and 570 draw the left- and right-hand sides respectively.

Strictly speaking, the size-determining variables, F1 through F4, need not be defined separately as is done in line 522. Line 555, for example, could be written as: 555 FORF=15553TO15615:POKEF,131:NEXT. This would certainly save variable space in the computer memory and speed up operations a little bit. In later projects, however, you will be changing the size of the border figure and, in fact, using its parameters as reference points for drawing other figures on the screen.

In the long run, defining the border in terms of variables F1 through F4 gives the subroutine a degree of usefulness and flexibility that would be lost otherwise.

The variables introduced by FIELD-1 are border parameters F1 through F4 and a drawing variable F. F serves its entire
purpose through this subroutine and can thus be redefined and
used for other purposes in other subroutines. Variables F1
through F4, however, ought to be reserved. The subroutine
requires no initializing values, so it can be tested and run
without having to set up any values ahead of time.

When you are ready to go, enter FIELD-1 into your computer using the designated line numbers. Test the routine with the
simple test program. The test program merely clears the screen
and calls FIELD-1. You know everything is in good working
order when you see a nice, clean rectangle drawn on the screen.
Line 2Ø in the test routine lets you clear the screen and redraw
the border any number of times by simply striking the ENTER
key.

Subroutine 6ØØ Initialize

For the purposes of this particular program, the 6ØØ IN-
ITIALIZE block in the flowchart takes the form of INITIAL-1 in
the programming.

For the time being, it is sufficient to say that this subroutine initializes the motion of the ALPHA-1 creature. The
values of variables X and Y determine the actual coordinates of
the creature on the screen—at position 1Ø, 1Ø in this case. The
values of I and J determine the velocity of the creature's motion,
with I being the horizontal component and J being the vertical
component of motion.

There will be much more explained about the position
coordinates, X and Y, and the velocity vectors, I and J, later in
this section. The simple purpose of the INITIAL-1 subroutine
would be lost in a detailed discussion at this point, though.

Here is the simple INITIAL-1 subroutine:

```
600 REM ** INITIAL-1 **
610 X=10:Y=10:I=I:J=1
620 RETURN
```

Variables introduced: X, Y, I, J
Variables required at start: none

One might argue that this little program is too simple to be
designated as a separate subroutine. That might be the case, but
since it is used so frequently throughout the work in this book, it
is given its own place as a subroutine, thereby making it
unnecessary to type it into the machine each time you begin
building up a new program.

725 CONTACT Conditional Subroutine

From the perspective of a flowchart analysis, the purpose of the CONTACT conditional is to determine whether or not the creature's next move is going to make it hit or jump over the border figure. The creature must be kept within the confines of its rectangle, and CONTACT is the scheme responsible for doing the job.

The CON SENSE-1 subroutine is one of several contact-sensing schemes to be used in this book. In this case, the system scans the path in front of the moving creature, looking for a lighted screen segment that indicates an obstacle in its path. In the ALPHA-I BASIC program, the only possible obstacle is the border drawing, and that's what the CON SENSE-1 subroutine looks for.

As in the case of the INITIAL-1 subroutine, it is difficult to describe the finer features of CON SENSE-1 until you have gone through a detailed explanation of how the creature moves on the screen. For the time being, you can look over the programming for the subroutine without worrying too much about how and why certain functions and statements are used. Lines 745 and 750 do contain some information you can appreciate at this time.

Whenever CON SENSE-1 picks up a potential contact with an obstacle, it sets variable CO equal to 1 before returning to the main program. If, on the other hand, the path ahead appears to be clear, CON SENSE-1 returns with CO equal to $\emptyset$. This feature is noted on both of the CONTACT conditional blocks on the flowchart in Fig. 3-2.

```
725 REM ** CON SENSE-1 **
730 FORXP=2TO2+ABS(I):FORYP=1TO1+ABS(J)
735 IFPOINT(X+SGN(I)*XP,Y+SGN(J)*YP=-1THEN75
740 NEXT:NEXT
745 CO=1:RETURN
750 CO=1:RETURN
```

Variables introduced: CO, XP, YP

Variables required to start: X, Y, I, J

650 FETCH NEW MOVE Subroutine

This is the operation that is responsible for selecting a new set of speed/direction motion vectors for the creature. Accord-

ing to the flowchart in Fig. 3-2, this subroutine is called only
when CONTACT senses an impending obstacle and generates a
value of CO equal to 1.

Here is the program version of this operation:

```
650 REM ** FETCH NEW-1 **
660 RI=RND(5):RJ=RND(5)
665 IFRI=3ANDRJ=3THEN660
670 RETURN
```

Variables introduced: RI, RJ
Variables required at start: none
Test routine
```
10 GOSUB650
20 PRINTRI,RJ
30 GOTO10
```

This subroutine doesn't require an extensive understanding
of the overall system to appreciate how it works. Line 660
simply generates two random numbers between 1 and 5. These
integer values are assigned to variables RI and RJ.

Line 665 is responsible for eliminating the condition where
both random integers are equal to 3. As shown a bit later in this
section, a pair of 3s would result in a stop code. They would
cause the creature to stop moving on the screen, and in the
context of the simple ALPHA-I BASIC system, this particular
option cannot be tolerated; the creature, in effect, would die on
the spot. Line 665 eliminates the stop code by cycling the
program back to 660 where it takes another crack at picking a
combination of numbers. Control returns to the main program
only if the values of RI and RJ are not *both* equal to 3.

If you load FETCH NEW-1 and the suggested test routine
into your system, entering RUN will cause an endless pair of
columns to appear on the screen. The columns will be composed
of integers between 1 and 5 inclusively, but any given pair
should not be equal to 3.

THE MAIN PROGRAM

The main program follows the flowchart in Fig. 3-2, calling
the necessary subroutines and finishing up the flowchart by
adding a few other operational steps. This program is named
ALPHA-1, BASIC-1 MASTER.

Line 1Ø calls subroutine 5ØØ, the heading/start subroutine. After the subroutine, the program clears the screen (line 15), draws the border figure (Subroutine 55Ø called by line 2Ø), and initializes the creature's position and motion (Subroutine 6ØØ called by line 3Ø).

```
5 REM ** ALPHA-I, BASIC-1 MASTER **
10 GOSUB5ØØ
15 CLS
20 GOSUB55Ø
30 GOSUB6ØØ
40 X=X+I:Y=Y+J
50 GOSUB725
60 IFCO=0THEN8Ø
65 SET(X,Y):SET(X+1,Y):RESET(X,Y):RESET(X+1,Y):GOSUB65Ø
70 I=RI-3:J=RJ-3:GOSUB725
75 IFCO=1THEN65
80 SET(X,Y):SET(X+1,Y):RESET(X,Y):RESET(X+1,Y(
85 GOTO4
```

Line 4Ø is responsible for setting up the next position for the creature. It is represented by the SET NEXT MOVE block on the flowchart.

It is really a rather simple set of statements. The X variable represents the horizontal position of the creature at any given moment, and I represents the horizontal component of the creature's motion. So if the creature's X position happens to be 2Ø and the I vector is +2, the first part of line 4Ø will set a new value of X at 2Ø+2, or 11. The Y variable, by the same token, represents the current vertical position of the little creature, and J is the vertical component of its motion. So if Y happens to be equal to 15 and J is −2, it computes the second half of line 4Ø to establish a new Y position of 15-2, or 13.

The integer values of X and Y can be anywhere between zero and some positive number that is dictated by the graphics format of the computer. In the case of Radio Shack's TRS-80, X values can be anywhere between 0 and 127, while the Y values can run anywhere between 0 and 47.

According to line 4Ø of the main program, the X and Y coordinates are actually determined by the values of vectors I and J. The larger these values happen to be, the faster the values of X and Y change. This means the larger the values of I and J, the faster the creature appears to move across the screen.

The range of values for vectors I and J are fixed at integers between −2 and +2 inclusively. This range is determined by our own programming, rather than by the nature of the computer.

Figure 3-3 shows all possible motion vectors that can be derived from this I, J format. The arrows indicate the direction of motion and the lengths of the arrows indicate the relative speed of motion. You can see that the creature has 24 different motion vectors at its disposal, 24 different combinations of directions and speeds. Having two sets of numbers between -2 and $+2$ actually provides 25 possible combinations, but the 0,0 combination is eliminated. Why is I=0, J=0 eliminated? Plug those valued for I and J into line 4Ø of the master program, and you will see that there is no change in the creature's position. A 0,0 vector combination causes the creature to stand motionless on the screen. That's why the combination is eliminated from the drawing in Fig. 3-3.

Recall that the INITIAL-1 subroutine sets the values of I and J at 1. The creature thus begins his life on the screen by moving in the 1,1 direction, and you can see from Fig. 3-3 that this combination carries the creature at an angle downward and to the right at a moderate speed.

If you are accustomed to working with X-Y graphs in the context of engineering and analytic geometry, you might be disturbed by a couple of secondary features of the graph in Fig. 3-3. For one thing, the positive and negative senses of the vertical components are reversed from the conventional mode. This is necessary in this case because the TRS-80 screen format calls for moving downward by increasing the values of the Y coordinate, and to move upward, the values of Y must be decreased. Hence, the need for having $-J$ toward the top of the screen and $+J$ toward the bottom. The horizontal components of motion follow the usual convention, with positive motion being toward the right and negative motion toward the left.

Another little feature is the fact that the TRS-80 screen format increments different distances for horizontal and vertical motion. An incremental step in the horizontal direction, in fact, is only half the distance covered by an incremental step in the vertical direction. In the conventional format, vectors having equal values causes motion at a 45-degree angle. But here, 45-degree motion is achieved only by having the I component equal to twice the value of the J component.

Actually, the creature doesn't care a whit about the sense of its vertical component of motion or the vertical stretching of its geometry. You can distort the surface in any way your imagination allows, and these little creatures would adapt to it with no problem at all. Robot schemes built around position-sensing

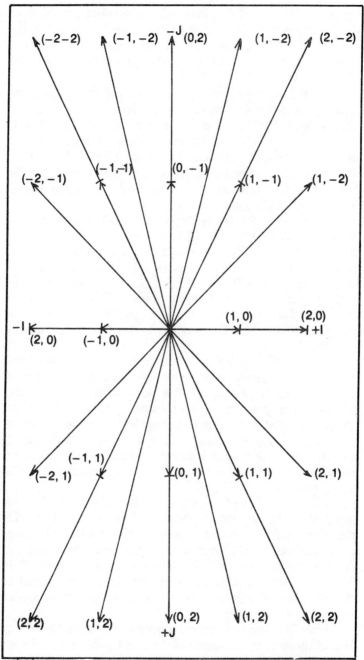

Fig. 3-3. Vector chart for creature motions.

formats would blow their minds with the simplest distortions of their environment. In fact, the only reason to bring up the subject of distorted fields is that the information is helpful later on when it is time to do a statistical analysis of how the ALPHA creature behaves.

All of this merely justifies step 4Ø in the master program. This line establishes a new position on the screen by taking the old position and adding a signed integer to the X and Y values. It is important to realize that the creature does not immediately go to the new position determined by line 40, SET NEXT MOVE. Rather, it is a suggested position that must be tested for possible contact with an obstacle (namely the border figure) before actually taking that next step.

Here's how the tricky little CON SENSE-1 subroutine works. The SGN(I) and SGN(J) functions in line 735 pick up the direction of motion. Remember that the direction is determined by the signs of the I and J values. The speed of motion is determined by ABS(I) and ABS(J) functions in line 73Ø. The proposed values of X and Y appear in line 735, and you ought to remember that they have already been tentatively established by the SET NEXT MOVE operation in line 4Ø of the master program.

So CON SENSE has the creature sitting in a hypothetical position on the screen, and it knows both the sense and speed of both components of motion. The question is this: Has the proposed new position for the creature carried it up to an obstacle, carried it right into an obstacle, or carried it over an obstacle? If any one of these conditions exist, CON SENSE-1 must respond with a signal indicating a contact situation. If none of the three conditions is met, the proposed path for motion is free of obstacles.

CON SENSE-1 effectively scans the area in front of its direction of motion, and the number of spaces it inspects depends on the rate of motion. The faster the creature is moving, the farther out it must "look" for barriers. Line 73Ø sets up the criteria for determining how far ahead the system searches. This is why the absolute values of the motion codes are relevant in that line, the larger these values become, the greater the size of the area to be inspected.

Line 735 then determines, among other things, the direction of the area to be inspected. Of course, the most relevant area to be checked is that directly ahead of the creature's path of motion, and this is why the signs of the I and J components are relevant in the line. Line 735 also incorporates a POINT function which, in the Radio Shack BASIC, returns a −1 value if the coordinates of the point indicate a lighted spot on the screen. If the point is not lighted, the function returns a 0.

In essence, CON SENSE-1 scans the area in front of the path of the creature's motion looking for lighted spots representing parts of some kind of barrier. The scheme anticipates the situation, always working one step ahead of where the creature actually appears on the screen. When this contact-sensing routine is used properly, the creature never quite touches or jumps over any barrier in its path.

In passing, you might note that the scheme only checks as far ahead as necessary. It would be possible to simplify the subroutine by having it scan the maximum amount of area each time. The only problem with this idea is that the maximum area involves looking ahead at 9 positions, as opposed to an average of 6 actually required. The searching operation is a time consuming one that can slow down the action on the screen quite a bit—a very noticeable amount. By searching ahead no more than absolutely necessary, the whole ALPHA-I BASIC scheme runs about ½ faster that it would otherwise.

Calling subroutine 725 in line 5Ø of the master program causes the system to search the path immediately ahead. If a contact situation is impending, this subroutine returns with CO=1. Otherwise it returns to line 6Ø with CO=Ø.

If a contact is the next thing in order for the creature, line 65 of the master program causes the creature to blink on and off in its new position. This is done simply to let the experimenter know where the creature is residing on the screen.

BLINK in line 65 is accomplished by simply doing a set of two SET statements followed by a set of two corresponding RESET statement. The line could be simplified by doing SET(X,Y): RESET(X,Y), but the 2:1 distortion of the graphics field would make the creature appear as a small rectangle, rather than a neat little square. To make it appear as a neat little square, one merely lights up both the X and X+1 must both be RESET to erase the complete creature image.

After doing BLINK, the main program calls for fetching a new motion code, enter FETCH NEW-1. FETCH NEW-1 has already been described in some detail. It is the subroutine that selects pairs of random numbers between 1 and 5 and automatically excludes a 3, 3 combination. After seeing exactly how the creature moves, you are now in a better position to appreciate the purpose of FETCH NEW-1.

In effect, FETCH NEW-1 selects new motion vectors, but they come from this subroutine as RI and RJ, rather than I and J. They emerge as integers between 1 and 5, rather than integers between −2 and +2. However, examine the DECODE block, represented by line 70 in the master program. It subtracts 3 from both RI and RJ. The numbers between 1 and 5 are thus decoded into numbers between −2 and +2. The motion-stopping combination of I=J=0 never comes about because the FETCH RAND-1 is not allowed to return with RI=RJ=3.

This is another one of those operations that might seem to be needlessly complicated. Why not have FETCH NEW MOVE generate numbers already in the range of −2 to +2 thereby eliminating the need for a DECODE operation? It would be easier to do it that way for most of the ALPHA-class projects in this book. The higher-order machines, however, call for FETCH NEW MOVE numbers that are larger than zero. FETCH NEW-1 is going to be used in most of the programs in this book, so instead of writing different FETCH NEW MOVE subroutines again and again, you can use this single one wherever the operation is necessary.

At any rate, you now have a valid set of motion codes. The next question is whether or not the new set of codes will get the creature away from the barrier. The question is quite easily answered by applying the CONTACT subroutine a second time. Will the newly fetched motion codes get the creature out of trouble? If not, CO=1 again and the BLINK, FETCH, and DECODE sequence must be run all over again.

The MOVE operation (line 80) is identical to BLINK. The creature is in its new position by this time, however, and it is running free of obstacles in its path.

The overall impression created by this program and its subroutines is a rectangle that serves as sort of a playpen for a little square creature that moves around within its confines, blinking along in a straight-line path until it encounters one of the edges of the rectangle. It then blinks at that barrier until it picks up a new motion code that allows it to move along or away from that barrier.

The action is fairly fast for the ALPHA-I BASIC program, and for the uninitiated, it appears to be some sort of video game that has a ball rebounding inside a rectangular field.

LOADING AND TESTING THE PROGRAM

Load the subroutines in the order they are presented in this chapter. Where feasible, use the suggested test routines to check them out before loading the next one. Until you've had a chance to study the entire program in this book, it's a good idea to use the line numbers and subroutine names specified here. If you find you are having any trouble running the programs, check your program listing against the composite listing at the end of this chapter.

When everything is ready to go, the program is initiated by entering RUN. The heading messages should appear first and remain on the screen until you strike the ENTER key. After that, sit back and watch the little creature go through its Alpha-like behavior. The program will run indefinitely, until you either do a BREAK, reset the system, or turn it off.

Your appreciation for what is going on at this point is directly proportional to your imagination and understanding of how it all works. To be sure, there are far simpler programs for getting a square to bounce around inside a rectangle figure on the screen, but none of them contain the germ of machine intelligence that this one does.

Simply realizing that the basic amoeba dancing around in a drop of water has a germ of life which separates it from inert matter makes it something quite awsome in its own right. And in spite of its innocuous appearance, the ALPHA-I BASIC creature has the heart of a machine-creature of vast potential.

COMPOSITE ALPHA-I BASIC PROGRAM

```
5 REM ** ALPHA-I, BASIC-1 MASTER **

10 GOSUB500

15 CLS

20 GOSUB550

30 GOSUB600

40 X=X+I:Y=Y+J

50 GOSUB725

60 IFCO=0THEN80

65 SET(X,Y):SET(X+1,Y):RESET(X,Y):RESET(X+1,Y):GOSUB650

70 I=RI-3:J=RJ-3:GOSUB725

75 IFCO=1THEN65
```

```
80 SET(X,Y):SET(X+1,Y):RESET(X,Y):RESET(X+1,Y)
85 GOTO40
500 REM ** HEADING-1 **
505 CLS
510 PRINTTAB(25)"RODNEY ALPHA-i":PRINT
515 PRINTTAB(25)"BASIC VERSION"
520 FORL=0TO10:PRINT:NEXT
525 INPUT"DO ENTER TO START";S$:RETURN
550 REM ** FIELD-1 **
552 F1=15553:F2=15615:F3=16321:F4=16383
555 FORF=F1TOF2:POKEF,131:NEXT
560 FORF=F3TOF4:POKEF,176:NEXT
565 FORF=F1TOF4STEP64:POKEF,191:NEXT
570 FORF=F2TOF4STEP64:POKEF,191:NEXT
575 RETURN
600 REM ** INITIAL-1 **
610 X=10:Y=10:I=1:J=1
620 RETURN
650 REM ** FETCH NEW-1 **
660 RI=RND(5):RJ=RND(5)
665 IFRI=3ANDRJ=3THEN660
670 RETURN
725 REM ** CON SENSE-1 **
730 FORXP=2TO2+ABS(I):FORYP=1TO1+ABS(J)
735 IFPOINT(X+SGN(I)*XP,Y+SGN(J)*YP)=-1THEN750
740 NEXT:NEXT
745 CO=0:RETURN
750 CO=1:RETURN
```

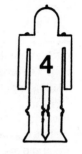

Alpha-I With Scoring

The ALPHA-I BASIC creature is certainly fun to watch especially if you know what is going on behind the scenes. Now is the time to introduce some science into the project, however. ALPHA-I with a scoring feature is still a lot of fun, maybe even more fun, but the idea is to be qualifying some of the creature behavior, giving you something of a handle on how successful the creature can be with operting in this ALPHA mode.

Figure 4-1 shows the screen format generated by the programming in this chapter. The rectangular border and creature figures are exactly the same as those used in ALPHA-I BASIC. In fact they are generated by the same subroutines.

In this case, however, note the scoring messages across the top of the screen. NO. CONTRACTS specifies the total number of times the creature encounters the barrier, including all the times it makes moves that don't work out. NO.GOOD MOVES keeps track of the number of moves that actually get the creature away from a contact situation. If you think about it for a moment, you ought to realize that the number of good moves is always less than or equal to the total number of contact situations.

SCORE is simply the ratio of number of good moves to the total number of contacts. It is an up-to-the-moment score of the creature's performance. The higher the score, the better the performance— the more successful the creature is at getting away from the barrier.

SOME THOUGHTS ABOUT ALPHA-I SCORING

Just how well can an ALPHA-I creature be expected to score within its typical rectangle figure? Well, the creature is selecting motion codes at random, and it is safe to assume the chances of picking one certain motion code is just about as good as selecting any other one.

Figure 3-3 in the previous chapter shows all possible motion vectors for the ALPHA-I creature. The creature programmed into the system in this chapter uses exactly the same set of motion codes, so the chance of picking any one of the 24 different motion codes is 1:24.

The creature is going to score much better than 1:24, or .042, simply because there is more than one motion code that can carry it away from any barrier situation. If, for instance, the creature runs into a section of the barrier that is some significant distance away from a corner, the chances of picking one of the good moves is between .0500 and .583, depending depending on the angle of contact.

To see how this works for yourself, lay a sheet of paper over the figure in Fig. 3-3, arranging it so the edge of the paper cuts through the origin or center of the graph. The exposed vectors represent the family of workable responses available to the creature. Rotate the paper around the origin of the graph and you will find between 12 and 14 possible motion codes depending on the angle of the paper's edge. (The edge of the paper, of course, represents a barrier). When doing the counting in this particular experiment, remember that the creature can slide along the edge of a barrier.

If the creature's barrier didn't have any corners, you would expect the score to run rather consistently between .5 and .58 over the long run. But of course, any enclosing 2-dimensional figure must have at least three places where the border changes directions. (The simplest figure would be a triangle, whereas a figure having an infinite number of places where the direction changes would be a circle.) For our purposes here the figure is a rectangle having four corners, and the question is this: What are the chances of picking a good motion code when encountering a corner?

To answer this question, fold a sheet of paper into fourths and cut out one of the ¼th sections. The right angle cut out of the paper represents a corner barrier. Place the paper over the drawing in Fig. 3-3 and arrange it so the corner barrier is at the origin of the graph. Rotate the paper through all the possible contact angles and, in each case, count the number of possible "escape" motion codes. In each case, you will find 4, 5, 6, or 8 workable motion code. This means the creature's chances of getting out of a corner trap is anywhere between 4:24 (.167) and 8:24 (.333).

It is possible to carry this sort of analysis to a rather lengthy and, perhaps, boring extreme. The bottom line, however, is that the ALPHA-I creature, working in the rectangular area specified here, will certainly score something less than .58, but better than .167. When you get the scoring program set up and running, you will find the creature scoring somewhere between .460 and .521 after a hundred contacts.

If you are interested in pursuing this matter of ALPHA-I scoring under different barrier conditions (varying the size of the rectangle, changing to a triangular or circular field), and so on, you will find some suggested experiments at the end of this chapter. There is a lot of original thinking and experimentation that must be done in this particular area of adaptive machine intelligence.

ALPHA-I SCORING FLOWCHART

The flowchart for the ALPHA-I scoring program is shown in Fig. 4-2. Note that its general flow of operations is almost identical to the ALPHA-I BASIC flowchart in Fig. 3-2. And what's more, you can see that it uses many of the same subroutines.

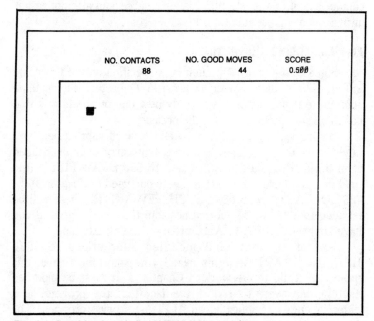

Fig. 4-1. Graphics format for ALPHA-I with scoring.

As far as the flowchart is concerned, the only differences are the addition of the following operational blocks:

```
C=C+1
UPDATE SCORE
D=D+1
```

C=C+1is a counting operation which keeps track of the total number of times the creature has encountered a barrier situation. Note that the C=C+1 block is situated so that it increments one count whenever either of the two CONTACT conditionals show a contact situation, CO=1. So every contact, whether from hitting a barrier while moving freely across the screen or from making a bad selection of a new move, results in a higher contact count. The D=D+1 block is responsible for counting the number of good selections after running into a barrier. Note that this block is situated at the point where the creature notes no contact *after* selecting a new motion code.

The output block, UPDATE SCORE, appears twice on the flowchart. Its job is to print the messages along the top of the screen and calculate the SCORE. The messages are updated whenever there is any change in the scoring figures, after each contact and after the creature makes a successful attempt to get away from a contact situation.

ALPHA-I SCORING SUBROUTINES

Some of the subroutine numbers specified on the flowchart in Fig. 4-2 are marked with an asterisk. The asterisk indicates a subroutine that is either an entirely new one or an old one that has to be slightly modified for the occasion.

Before looking at these new and revised subroutines, you might want to note the subroutines that carry over unchanged from the ALPHA-I BASIC program: S. 55Ø DRAW FIELD (the field in this program is identical to the one used in Chapter 3), S. 725 CONTACT, and S.65Ø FETCH NEW MOVE. There will be no need to enter these subroutines into the new program if you have kept the ALPHA-I BASIC program on cassette tape.

As for the new and modified subroutines: S. 5ØØ HEADING/START performs exactly the same function in this project as it did in the work in Chapter 3. It must be modified slightly, however, to reflect the fact that the program is a different one— specifically, an ALPHA-I scoring program. The file name in this instance is HEADING-2. Compare the program

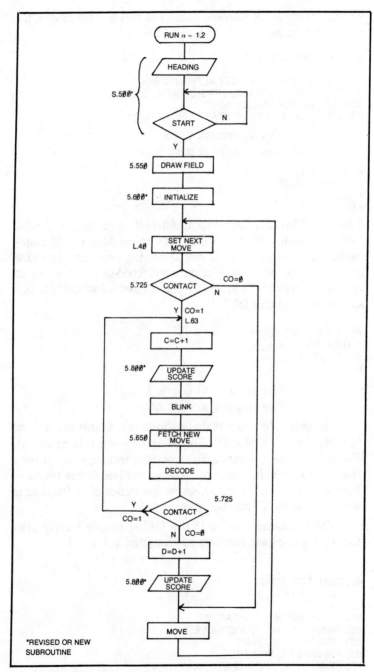

Fig. 4-2. Flowchart for ALPHA-I with scoring.

with HEADING-1 in Chapter 3, and you will be able to spot the minor differences.

```
500 REM**HEADING-2**
505 CLS
510 PRINTTAB (25) "RODNEY ALPHA-I":PRINT
515 PRINTTAB (20) "BASIC VERSION WITH SCORING"
520 FORL=0 T010 : PRINT: NEXT
525 INPUT "DO ENTER TO START";S$: RETURN
```

> Variables introduced: L, S$
> Variables required at start: none

Test routine
```
1 CLS:GOSUB500
2 CLS:PRINT"OK"
3 INPUTX$:GOTO1
```

S.600 INITIALIZE isn't much different from the ALPHA-I BASIC version, either. All that neds to be done is add statements for initializing the C and D scoring counters. The new line, line 615, sets these contact and good-move counters to zero at the start of the program. The revised S.600 INITIALIZE subroutine is labeled INITIAL-2.

```
600 REM ** INITIAL-2**
610 X=10:Y=10:I=1:J=
615 C=0:F=0
620 RETURN
```

> Variables introduced: C,Y, I, J, C, D
> Variables required at start: none

The only subroutine that is entirely new to this experiment is S.800 UPDATE SCORE. For filing purposes, it is named UD SCORE-1. Line 805 uses a field-specifier technique for printing the score figures in a uniform fashion, and line 810 is responsible for calculating the score, given the values of D (number of good moves) and C (number of contacts.)

The remaining lines in UD SCORE-1 merely merely print the scoring titles and numbers as illustrated in Fig. 4-1.

```
800 REM ** UD SCORE-1 **
805 U$="#.###"
810 E=D/C
850 PRINT@15, "NO. CONTACTS"
852 PRINT@35, "NO. GOOD MOVES"
853 PRINT@55, "SCORE"
855 PRINT@87, C:PRINT/108, D
860 PRINT@119, USINGU$;E
865 RETURN
```

Variables introduced: U$
Variables required at start: D, C

Test routine
```
1 CLS:INPUT"C, D"; C,D: GOSUB8ØØ
2 INPUTX$:GOTO1
```

THE MASTER PROGRAM FOR ALPHA-I SCORING

Just as the flowchart for ALPHA-I scoring is practically identical to the chart for ALPHA-I BASIC, it follows that the master programs are almost the same. The only differences of significance are the statements in lines 63 and 78 which call for incrementing the counters and updating the scoring display.

Comparing the two master programs, you'll find you can create the new one by altering the REM statement in line 5 and then adding lines 63 and 78.

```
5 REM ** ALPHA-I, BASIC-2 MASTER**
10 GOSUB5ØØ
15 CLS
20 GOSUB55Ø
30 GOSUB6ØØ
40 X=X+I:Y=Y+J
50 GOSUB725
60 IFCO=ØTHEN8Ø
63 C=C+1:GOSUB8ØØ
65 SET(X, Y): SET (X+1C Y):RESET (X,Y): RESET (X+1C Y):GOSUB650
70 I=RI−3:J=RJ−3:GOSUB725
75 IFCO = 1 THEN63
78 D=D+1:GOSUB8ØØ
80 SET (X,Y):SET (X+1,Y):RESET (X,Y):RESET (X+1,Y)
85 GOTO4Ø
```

LOADING THE ALPHA-I SCORING PROGRAM

If the ALPHA-I BASIC programming is already resident in your computer's program memory, you should have no real problem creating the scoring version from it. Just do this:

1. Edit S.5ØØ HEADING/START programming so that it lines up with the HEADING-2 subroutine.

2. Edit S. 6ØØ INITIALIZE so that it has the initializing steps for C and D as shown in INITIAL-2.

3. Add S.8ØØ as UD SCORE-1.

4. Edit the master program by altering line 5 and adding lines 63 and 78.

Once you've had a chance to run this program and make certain it is working as it should, save the program on cassette tape. It will be used as the foundation for creating other programs later on. If you have any difficulties with the operation of the program, compare your listing with the composite program at the end of this chapter.

SUGGESTED EXPERIMENTS FOR ALPHA-I SCORING

There are a number of relevant questions about the performance of simple ALPHA-I creatures that must be answered and proven. For example, how is the scoring affected by changing the size and aspect ratio of the rectangle? How would using a triangular barrier effect the creature's scoring? What is the relationship between the creature's size and the size of its border figure, as reflected in the score?

The scoring technique suggested in this chapter is far more sensitive to changes when the total number of contacts is relatively low than when the number of contacts is high. Is there an alternate method for scoring the quality of the creature's choices so that the score is equally sensitive throughout the experiment?

Suppose you place a rectangular figure inside the main field rectangle and make the creature sensitive to both figures. You then rewrite the program so that no counting or scoring takes place when the creature encounters the outer border, but the counting and scoring do take place when the creature stumbles into the inner rectangle. How will the scoring be different from that described in this chapter?

Try building a reasonable facsimile of a circle on the screen and use it as the border figure. Is the scoring different from that found for the usual rectangular border?

COMPOSITE PROGRAM FOR ALPHA-I WITH SCORING

```
5 REM ** ALPHA-I, BASIC-2 MASTER **

10 GOSUB500

15 CLS

20 GOSUB550

30 GOSUB600

40 X=X+I:Y=Y+J

50 GOSUB725

60 IFCO=0THEN80

63 C=C+1:GOSUB800
```

```
65 SET(X,Y):SET(X+1,Y):RESET(X,Y):RESET(X+1,Y):GOSUB650

70 I=RI-3:J=RJ-3:GOSUB725

75 IFCO=1THEN63

78 D=D+1:GOSUB800

80 SET(X,Y):SET(X+1,Y):RESET(X,Y):RESET(X+1,Y)

85 GOTO40

500 REM ** HEADING-2 **

505 CLS

510 PRINTTAB(25)"RODNEY ALPHA-¥":PRINT

515 PRINTTAB(20)"BASIC VERSION WITH SCORING"

520 FORL=0TO10:PRINT:NEXT

525 INPUT"DO ENTER TO START";S$:RETURN

550 REM ** FIELD-1 **

552 F1=15553:F2=15615:F3=16321:F4=16383

555 FORF=F1TOF2:POKEF,131:NEXT

560 FORF=F3TOF4:POKEF,176:NEXT

565 FORF=F1TOF4STEP64:POKEF,191:NEXT

570 FORF=F2TOF4STEP64:POKEF,191:NEXT

575 RETURN

600 REM ** INITIAL-2 **
610 X=10:Y=10:I=1:J=1

615 C=0:D=0

620 RETURN

650 REM ** FETCH NEW-1 **

660 RI=RND(5):RJ=RND(5)

665 IFRI=3ANDRJ=3THEN660

670 RETURN

725 REM ** CON SENSE-1 **

730 FORXP=2TO2+ABS(I):FORYP=1TO1+ABS(J)

735 IFPOINT(X+SGN(I)*XP,Y+SGN(J)*YP)=-1THEN750

740 NEXT:NEXT

745 CO=0:RETURN
```

59

```
750 CO=1:RETURN
800 REM ** UD SCORE-1 **
805 U$="$.###"
810 E=D/C
850 PRINT@15,"NO. CONTACTS"
852 PRINT@35,"NO. GOOD MOVES"
853 PRINT@55,"SCORE"
855 PRINT@87,C:PRINT@108,D
860 PRINT@119,USINGU$;E
865 RETURN
```

Fun And Games With Alpha-I

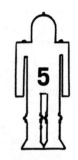

The fun and excitement of working with adaptive intelligence is not entirely limited to intellectual fun and excitement. It is quite possible to play around with the little critters, putting them into tricky situations they must cope with in their own unique fashion. Once you have entered and tested the ALPHA-I with scoring program, you are in a position to devise some games. The two described in this chapter are ALPHA-I MAZE and ALPHA-I PAINT INTO A CORNER.

The primary purpose is to show that it is possible to play with an ALPHA-I creature and derive some pleasure from watching it at work. Even so, you will find some principles introduced in this chapter that will have some special significance later on in the project.

ALPHA-I MAZE GAME

Whenever two or more computer buffs get together and begin discussing robots and artificial intelligence the subject of maze-solving seems to crop up, sooner or later, anyway.

The ALPHA-I creature isn't very good at solving mazes if you think in terms of a robot "learning" to solve a maze perfectly. An ALPHA-I can work through a maze situation in its own fashion, however. It's possible to make a game out of the problem by seeing how few contacts are required for doing the job. ALPHA-I solves the maze purely by chance, and there is no way to tell ahead of time how long the job will take. The creature isn't goal-oriented. So in effect, the critter is not trying to solve the maze in the first place, but he gets there, and that's really what counts in the long run.

ALPHA-I MAZE is built right into the ALPHA-I scoring program you should have loaded and saved from Chapter 4. All that's needed to get started with MAZE is to modify S.500

HEADING a little bit, add a DRAW MAZE subroutine, and add a few more lines to the master program.

Figure 5-1 is the graphics for ALPHA-I MAZE. It looks very much like the graphics for ALPHA-I with scoring, having only some additional maze lines and a goal position designated G. The ALPHA-I creature treats the five maze lines as though they are border lines, rebounding from them in a random fashion.

The creature begins the game in the upper left-hand corner of the maze (as dictated by S.600 INITIALIZE) and gradually works its way toward the goal. The scorekeeping features from ALPHA-I with score keep track of the number of contacts, number of good moves, and the creature's running score.

Upon reaching the goal position in the lower left-hand corner of the screen, the game comes to a stop and the message HIT !! appears in the message space in the upper left-hand corner of the screen. The maze game can then be restarted by simply striking the ENTER key.

Two or more players can make a game of the thing by taking turns restarting the maze problem. After each player has a chance to watch his creature run the maze, the player whose creature has the fewest number of contacts wins. In the case of a tie, the creature showing the better overall score is the winner.

The flowchart for ALPHA-I MAZE is shown in Fig. 5-2. It is a rather extensive flowchart, but assuming you've done your homework on earlier flowcharts, there isn't much to it. In fact, the flowchart shown here is identical to the chart for ALPHA-I with scoring (Fig. 4-2) with only a couple of exceptions. For one, the MAZE flowchart must include a block that is responsible for adding the maze lines to graphics format. This block appears as S.580 DRAW MAZE in Fig. 5-2.

Then too, there is a need for sensing when the creature reaches the goal space and also to give the player a chance to restart the game from scratch. All of this is handled by the conditionals GOAL and AGAIN and the output block GOAL MESSAGE. DRAW MAZE, of course, is inserted into the program very close to the beginning. In this instance it falls between the usual S.500 DRAW FIELD and S.600 INITIALIZE subroutines.

The goal sensing and restarting operations are inserted in the operational loop leading from MOVE to either SET NEXT MOVE or DRAW FIELD. Whenever the creature is *not* in the

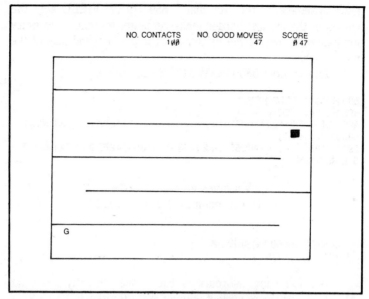

NO. CONTACTS   NO. GOOD MOVES   SCORE
1ØØ   47   Ø 47

G

Fig. 5-1. Graphics format for the ALPHA-I MAZE game.

goal area, the GOAL conditional responds with an N condition which directs operations back to SET NEXT MOVE.

If, on the other hand, the creature has just entered the goal area, GOAL responds with a Y and the GOAL MESSAGE is printed out on the screen. AGAIN brings all operations to a halt until it is satisfied (Y condition) by striking the ENTER key. Upon striking the ENTER key, the AGAIN conditional sends operations back up to the point where the field and maze are redrawn and the whole business is initialized again.

Subroutine Changes for ALPHA-I MAZE

As far as the subroutines are concerned, you only have to modify S.5ØØ HEADING to reflect the fact that this is a MAZE game rather than the standard ALPHA/I with score. Here is the revised version of this particular subroutine.

```
500 REM ** MAZE HEADING **
505 CLS
510 PRINTTAB (25) "RODNEY ALPHA-I":PRINT
515 PRINTTAB (25) "MAZE GAME"
520 FORL=ØTO1Ø:PRINT:NEXT
525 INPUT 'DO ENTER TO START";S$:RETURN
```

The maze-drawing subroutine, S.58Ø DRAW MAZE, is an entirely new one. It turns out to be a fairly simple one if you take advantage of the fact that S.55Ø DRAW FIELD sets the values of the critical border measurements, the ones necessary for fixing the size of the border and the position and size of the maze lines.

Here is the S.58Ø DRAW MAZE subroutine:

```
580 REM ** MAZE-1 **
585 FORN=ØTO2
586 FORF=F1+129+N*256TOF2+12Ø+N*256:POKEF,176:NEXT:NEXT
590 FORN=ØTO1
591 FORF=F1+265+N*256TOF2+255+N*256:POKEF,176:NEXT:NEXT
595 RETURN
```

Variables introduced: N
Variables required at start: F1, F2

```
1 CLS:GOSUB550:GOSUB58Ø
2 INPUTX$:GOTO1
```

The following subroutines carry over directly from the ALPHA-I with score project without any alterations:

```
S. 550 DRAW FIELD (FIELD-1)
S. 600 INITIALIZE (INITIAL-2)
S. 800 UPDATE SCORE (UD SCORE-1)
S. 650 FETCH NEW MOVE (FETCH NEW-1)
S.725 CONTACT (CON SENSE-1)
```

Main Program for MAZE

The main program for the MAZE game is a modified version of the ALPHA-I with scoring program. Line 5 is edited to show it is the ALPHA-I MAZE program, and line 25 is inserted to call the DRAW MAZE subroutine. Everything else is the same down to line 85. Instead of doing a GOTO4Ø as is done in Chapter 4, it is replaced with a conditional logic statement representing the GOAL conditional on the flowchart in Fig. 5-2. Line 9Ø is added to do the GOAL MESSAGE and AGAIN operations.

The ALPHA-I MAZE program is rather easily built around the ALPHA-I with scoring program described in Chapter 4. If you have saved that program on cassettte tape, all you have to do now is load it into the computer and make the modifications and additions outlined here.

```
5 REM ** ALPHA-I MAZE **
10 GOSUB5ØØ
15 CLS
```

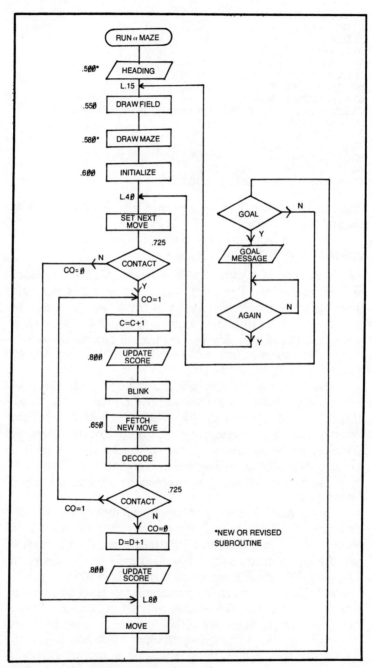

Fig. 5-2. Flowchart for the MAZE game.

```
20 GOSUB55Ø
25 GOSUB58Ø
30 GOSUB6ØØ
35 POKE16261,71
40 X=X+I:Y=Y+J
50 GOSUB725
60 IFCO=ØTHEN8Ø
63 C=C+1:GOSUB8ØØ
65 SET(X,Y):SET(X+1,Y):RESET(X,Y):RESET(X+1,Y):GOSUB65Ø
70 I=RI−3:J=RJ−3:GOSUB725
75 IFCO=1THEN63
78 D=D+1:GOSUB8ØØ
80 SET(X,Y):SET(X+(X+1,Y):RESET(X,Y):RESET(X+1,Y)
85 IFNOT(X<2ØANDY>41)THEN4
90 PRINT Ø,"GOAL !!":INPUTX$:GOTO15
```

The composite program listing for ALPHA-I MAZE appears at
the end of this chapter.

ALPHA-I PAINT INTO A CORNER GAME

Now here is a game that really puts ALPHA-I creature
abilities to the test. Personally, I have yet to tire of this little
game, and the people who see it in action get just about as
wound up in it as I do. The game is based on the old idea of
some thoughtless individual painting himself into the corner of a
room. The idea is to find some way to get out of the situation
without crossing any of the wet paint.

The "wet paint" in this case is a white trail left behind by
the ALPHA creature as it moves around on the screen. Recall
that S.725 CONTACT works by scanning the path ahead of the
creature, looking for segments of light which represent an
obstacle. So far in this book, the contact-sensing feature has
been applied only to sensing the creature's border figure. But
now, the contact subroutine will respond to the trail left behind
by the creature itself.

As the ALPHA creature moves around the screen, it reacts
in a random fashion to both the border figure and the trail it
leaves behind. Inevitably the little fellow will "paint" himself
into a trap. He then searches around at random, looking for an
unlighted path which leads to freedom. This is where you find
yourself starting to cheer the creature along. Move this way or
that! No, don't go that way—things will only get worse!

The creature loses every time. Sooner or later he paints
himself into a tiny rectangle of darkness which has no escape.
The game then comes to an end, and a special statement
displays the total number of "brush strokes" it was able to make.

To make a game of it, two more players can run a cycle of the game. The player whose creature accumulated the largest number of brush strokes wins the game.

Paint Into A Corner Flowchart and Subroutines

The flowchart for this PAINT game is quite similar to that of the ALPHA-I with scoring program in Chapter 4. All you will have to do is modify S.500 HEADING a little bit and add a few more statements to the master program.

Since it is assumed you have been studying the flowcharts thoroughly as you go along, it is now necessary to make a few comments about the additional blocks appearing in Fig. 5-3. The BS variable appearing in a couple of places on this flowchart is the brush-stroke counter. It is set to zero at the beginning of a game cycle, directly after the usual S.600 INITIALIZE subroutine. It is then incremented by one every time the creature makes a successful move. See the BS=BS+1 block following the MOVE step. The brush stroke score is not printed out, however, until the end of the game. See the PRINT DONE, BS output block on the flowchart.

Another new variable, CC, is introduced in this game. The purpose of CC is to count the number of successive moves that don't work. It is the criteria for ending the game. CC is set to zero whenever the system first senses contact with either the border figure or the creature's trail. CC is then incremented by CC=CC+1 as long as the creature is stuck in a trap and trying to bang its way out. Each unsuccessful attempt causes CC to grow larger. If the creature happens to find its way out of the situation before CC reaches 50, the subsequent contact resets CC to zero, and the creature has 50 more chances in succession to get away. The conditional statement, CC =50, is satisfied when the creature is finally down for the count. The program then prints the message DONE IN *BS* STROKES, where BS is the number of accumulated brush strokes.

The game comes to a halt at the AGAIN conditional and remains stopped until the player strikes the ENTER key. Hitting the ENTER key at this point in the operation satisfies the AGAIN conditional, and the program is restarted at S.550 DRAW FIELD.

PAINT INTO A CORNER does not call for any new subroutines. All the extra operations are covered in the main program. Only the S.500 HEADING routine has to be modified to show that it is the PAINT game.

So load ALPHA-I with scoring from your cassette machine, and fix up S.500 to look like this:

```
500 REM ** HEADING ALPHA PAINT **
505 CLS
510 PRINTTAB(25)"RODNEY ALPHA-I":PRINT
515 PRINTTAB(23)"PAINT INTO A CORNER"
520 FORL=0TO10:PRINT:NEXT
525 INPUT"DO ENTER TO START";S$:RETURN
```

Main Program for PAINT

The main program in this case is a slight variation of the main program for ALPHA-I with scoring. You can build around that earlier program and get the PAINT game going in short order.

```
5 REM ** ALPHA-I PAINT **
10 GOSUB500
15 CLS
20 GOSUB550
30 GOSUB600
35 BS=0
40 X=X+I:Y=Y+J
50 GOSUB725
60 IFCO=0THEN80
62 CC=0
63 C=C+1:GOSUB800
65 SET(X,Y):SET(X+1,Y):GOSUB650
70 I=RI-3:J=RJ-3:GOSUB725
75 IFCO=1GOTO90
78 D=D+1:GOSUB800
80 SET(X,Y):SET(X+1,Y)
85 BS=BS+1:GOTO40
90 CC=CC+1
95 IFCC>=50 THEN100 ELSE63
100 PRINT@0,"DONEIN":PRINT@64,BS-50'STROKES"
110 INPUTX$;:GOTO15
```

Line 35 is inserted to initialize the BS counter, and line 62 is added to set the CC counter to zero. The BS counter is then incremented by editing line 85 to include BS=BS+1. Line 90is added to increment the CC counter. Line 95 establishes the criteria for ending the game (50 bad moves in succession), with line 100 responsible for printing out the brush stroke score. Adding line 110 completes the operations by doing the AGAIN conditional operation. The composite program for this PAINT game appears at the end of this chapter.

THE ROBOT LAWNMOWER

This isn't really a game. Rather it is a demonstration of how an Alpha-Class machine can eventually cover all the space within

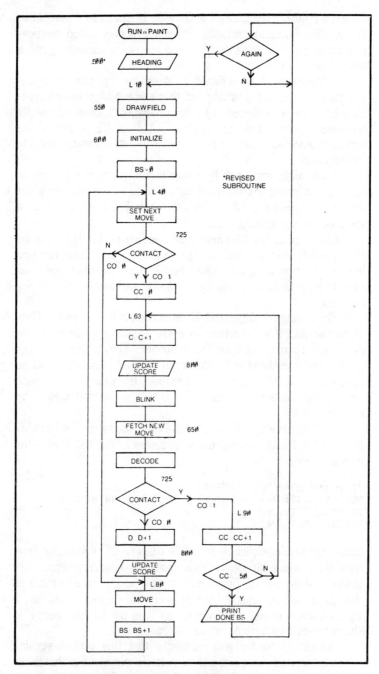

Fig. 5-3. Flowchart for PAINT INTO A CORNER.

a closed figure, even though its motion is essentially random in nature. The situation actually has more sophisticated implications, but it's more fun to work in the context of something like a lawn-mowing machine.

Incidentally, it is difficult to describe this particular demonstration without pointing out that some highly touted commercial machines claimed to be lawn-mowing "robots" are really no more sophisticated than our basic ALPHA-I creature — the simplest possible creature on the scale of evolutionary machine intelligence.

What might appear to be a significant achievement as far as the popular media is concerned turns out to be the starting point for what you're doing here. Does that help you see the power of the tool you are playing with?

At any rate, the "lawnmowing" version of ALPHA-I can be rather easily built around ALPHA-I with scoring. Assuming you have saved that program from Chapter 4 on cassette tape, you merely have to load it into the computer and make a few minor changes.

The changes suggested here accomplish one task: They allow the ALPHA-I creature to leave behind a white trail that does not stimulate contact-like behavior. The critter, in other words, can run through its own trail without regarding it as an obstacle. This, of course, is a turnabout from the PAINT game (where the creature did regard its own trails as obstacles to be reckoned with).

All you have to do is write one new subroutine and modify the ALPHA-I with scoring master program a little bit. Here's the new subroutine:

```
700 REM ** BORD POS SENSE-1 **
705 IF(X+I)>124OR(X+I)<4OR(Y+J)>45OR(Y+J)< 1ØTHEN715
710 CO=0:RETURN
715 CO=1:RETURN
```

This subroutine replaces the 725 CONTACT subroutine you have been using all along. This contact subroutine senses the position of the border figure and causes the creature to react as though it actually made contact with that figure. It is not a light-sensing contact mechanism, so it is totally insensitive to the creature's own tracks on the screen.

An astute reader will recognize that this contact-sensing scheme violates the principle of relative positioning that is so crucial to the success of adaptive perception. It's a small price to pay for a fun demonstration, however. An ALPHA-II machine described in a later chapter will do the same job within the proper context of adaptive perception. At that time there will be no need for adjusting the creature's behavior on the basis of its position relative to the geometry of its environment.

So enter S.700 CONTACT and then go through your master program for ALPHA-I with scoring, replacing all GOSUB725 statements with GOSUB700. Next, go through the master program, eliminating all statement sequences RESET(X,Y):RESET(X+1,Y). These are the statements used for erasing the trail as the creature moves along. You want the trails to remain so that you can clearly see how much territory the critter has covered at any given moment. Rename the master program something like LAWNMOWER and put it to work.

It's going to take a long time for the creature to cover all the space. For a long time, there are going to be some black spaces where the "lawnmower" hasn't gone. But rest assured, the ALPHA-I creature will cut all the grass. It might take a day or two, but the job will get done eventually.

Maybe you don't want to tie up the computer that long. That's all right. Just run the program long enough to get the idea of the thing.

COMPOSITE PROGRAMS

```
5 REM ** ALPHA-I MAZE  **

10 GOSUB500

15 CLS

20 GOSUB550

25 GOSUB580

30 GOSUB600

35 POKE16261,71

40 X=X+I:Y=Y+J

50 GOSUB725

60 IFCO=0THEN80

63 C=C+1:GOSUB800

65 SET(X,Y):SET(X+1,Y):RESET(X,Y):RESET(X+1,Y):GOSUB650

70 I=RI-3:J=RJ-3:GOSUB725

75 IFCO=1THEN63
```

```
78 D=D+1:GOSUB800

80 SET(X,Y):SET(X+1,Y):RESET(X,Y):RESET(X+1,Y)

85 IFNOT(X<20ANDY>41)THEN40

90 PRINT@0,"GOAL !!":INPUTX$:GOTO15

500 REM ** MAZE HEADING **

505 CLS

510 PRINTTAB(25)"RODNEY ALPHA-i":PRINT

515 PRINTTAB(25)"MAZE GAME"

520 FORL=0TO10:PRINT:NEXT

525 INPUT"DO ENTER TO START";S$:RETURN

550 REM ** FIELD-1 **

552 F1=15553:F2=15615:F3=16321:F4=16383

555 FORF=F1TOF2:POKEF,131:NEXT

560 FORF=F3TOF4:POKEF,176:NEXT

565 FORF=F1TOF4STEP64:POKEF,191:NEXT

570 FORF=F2TOF4STEP64:POKEF,191:NEXT

575 RETURN

580 REM ** MAZE-1 **

585 FORN=0TO2

586 FORF=F1+129+N*256TOF2+120+N*256:POKEF,176:NEXT:NEXT

590 FORN=00TO1

591 FORF=F1+265+N*256TOF2+255+N*256:POKEF,176:NEXT:NEXT

595 RETURN

600 REM ** INITIAL-2 **

610 X=10:Y=10:I=1:J=1

615 C=0:D=0

620 RETURN

650 REM ** FETCH NEW-1 **

660 RI=RND(5):RJ=RND(5)

665 IFRI=3ANDRJ=3THEN660

670 RETURN
```

72

```
725 REM ** CON SENSE-1 **
730 FORXP=2TO2+ABS(I):FORYP=1TO1+ABS(J)
735 IFPOINT(X+SGN(I)*XP,Y+SGN(J)*YP)=-1THEN750
740 NEXT:NEXT
745 CO=0:RETURN
750 CO=1:RETURN
800 REM ** UD SCORE-1 **
805 U$="#.###"
810 E=D/C
850 PRINT@15,"NO. CONTACTS"
852 PRINT@35,"NO. GOOD MOVES"
853 PRINT@55,"SCORE"
855 PRINT@87,C:PRINT@108,D
860 PRINT@119,USINGU$;E
865 RETURN
```

PAINT INTO A CORNER Program

```
5 REM ** ALPHA-1 PAINT **
10 GOSUB500
15 CLS
20 GOSUB550
30 GOSUB600
35 BS=0
40 X=X+I:Y=Y+J
50 GOSUB725
60 IFCO=0THEN80
62 CC=0
63 C=C+1:GOSUB800
65 SET(X,Y):SET(X+1,Y):GOSUB650
70 I=RI-3:J=RJ-3:GOSUB725
75 IFCO=1GOTO90
78 D=D+1:GOSUB800
```

73

```
 80 SET(X,Y):SET(X+1,Y)

 85 BS=BS+1:GOTO40

 90 CC=CC+1

 95 IFCC>=50THEN100ELSE63

100 PRINT@0,"DONE IN":PRINT@64,BS-50"STROKES"

110 INPUTX$;:GOTO10

500 REM ** HEADING ALPHA PAINT **

505 CLS

510 PRINTTAB(25)"RODNEY ALPHA-i":PRINT

515 PRINTTAB(23)"PAINT INTO A CORNER"

520 FORL=0TO10:PRINT:NEXT

525 INPUT"DO ENTER TO START";S$:RETURN

550 REM ** FIELD-1 **

552 F1=15553:F2=15615:F3=16321:F4=16383

555 FORF=F1TOF2:POKEF,131:NEXT

560 FORF=F3TOF4:POKEF,176:NEXT

565 FORF=F1TOF4STEP64:POKEF,191:NEXT

570 FORF=F2TOF4STEP64:POKEF,191:NEXT

575 RETURN

600 REM ** INITIAL-2 **

610 X=10:Y=10:I=1:J=1

615 C=0:D=0

620 RETURN

650 REM ** FETCH NEW-1 **

660 RI=RND(5):RJ=RND(5)

665 IFRI=3ANDRJ=3THEN660

670 RETURN

725 REM ** CON SENSE-1 **

730 FORXP=2TO2+ABS(I):FORYP=1TO1+ABS(J)

735 IFPOINT(X+SGN(I)*XP,Y+SGN(J)*YP)=-1THEN750

740 NEXT:NEXT
```

74

```
745 CO=0:RETURN
750 CO=1:RETURN
800 REM ** UD SCORE-1 **
805 U$="#.###"
810 E=D/C
850 PRINT@15,"NO. CONTACTS"
852 PRINT@35,"NO. GOOD MOVES"
853 PRINT@55,"SCORE"
855 PRINT@87,C:PRINT@108,D
860 PRINT@119,USINGU$;E
865 RETURN
```

6

Compiling Data
For Alpha-I Behavior

While it is fun to make up games for ALPHA-I, modify the programs to get it doing different tricks, and just watch it play around on the screen, don't lose sight of the fact that you're dealing with a system of some scientific and technological value. The science of the matter demands some unbiased figures regarding the behavior of your ALPHA-I creature. It is time to gather some data which you can use as a standard for comparing this creature's performance with that of more sophisticated versions you will be creating later on. This standard of comparison is absolutely essential if you hope to prove to yourself, and others, that something truly meaningful is taking place in front of you.

The measuring stick — the standard for comparison — used for demonstrating machine intelligence in this project is a simple sort of learning curve. It is a curve compiled from a large number of separate experiments, and it clearly reflects the creature's ability to adapt to its surroundings.

THE LEARNING CURVE FOR ADAPTIVE MACHINE INTELLIGENCE

Figure 6-1 is the learning curve for the basic ALPHA-I creature. In this particular case, the curve is the result of running 100 different ALPHA-I creatures through a cycle of 100 contacts each. Using the ALPHA-I program suggested in this chapter and running a hundred different creatures, your own curve should not be a significantly different from this one.

It is important to compile the scores for ALPHA creatures in a statistical fashion because no two creatures undergo exactly the same scoring pattern. You have probably noticed, for instance, that some ALPHA-I creatures seem to score exceedingly well at first, running perfect scores of 1.000 through the first four or five encounters with the border figure. Others, however, run into a corner within the first few moves and show exceptionally poor scores from the start. Hence, the need for a statistical compilation of results for a relatively large number of ALPHA creatures.

The idea is to create a statistically "average" ALPHA creature, one whose scoring pattern clearly reflects the random nature of ALPHA behavior. While the scoring pattern for an individual ALPHA creature might bounce around quite a bit during its lifetime, the "averaged" creature ought to show a fairly smooth curve such as the one in Fig. 6-1.

The general procedure for compiling the learning curve is to first run an ALPHA creature, keeping track of the total number of contacts (C) and good moves (D) at intervals of ten contacts. After running the creature through a series of 100 contacts and recording the data, the little fellow is wiped out and the whole procedure is started again.

The C and D scores, at intervals of 10 contacts, are then added to the same scores compiled through all previous trials. The result, after running maybe 100 different creatures, is a set of ten accumulated values for C and D. These numbers will be rather large because they represent the accumulated results of running 100 creatures.

The ten sets of C and D values are then used for computing the average score at each of the 10-step intervals. The result is

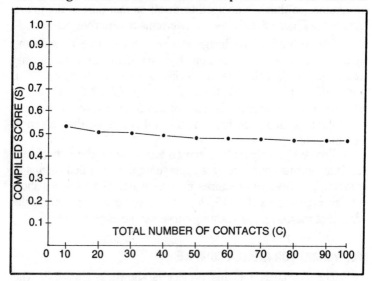

Fig. 6-1. A compiled learning curve for the basic ALPHA-I creature.

an average score for a large number of creatures after 10, 20, 30, ..., 100 contacts. Plotting this data on a graph results in a curve similar to that in Fig. 6-1.

Now it might sound like a lot of work to sit around watching the scores and writing down the results at 10-contact intervals for a hundred or more ALPHA-I creatures; it would be a lot of work to do the job by hand. Fortunately, you have a nice computing machine fully capable of doing all the tedious work for you. In fact, the compiling program suggested in this chapter makes the process fully automatic. All you do is enter the number of creatures to participate in the project, strike the ENTER key, and then go fishing or something like that for the rest of the day.

When the data-gathering part of the job is done, the system clears the screen and prints out the data necessary for drawing the learning curve. The program can be extended to draw the learning curve, itself. That job is left to you, however.

ALPHA-I COMPILE SCREEN FORMATS

The program for compiling the ALPHA-I learning curve involves three different screen formats: the heading, the creature and its environment, and the tabulated results. The heading format isn't much different from headings used in all the previous programs, so there is no need to dwell on that part of the job at this time. Figure 6-2 shows the two remaining screen formats.

Figure 6-2A is the image you see on the screen while the experiment is actually running. It is basically the same format used for the ALPHA-I with scoring project in Chapter 4. The only difference is the additional parameter, CREATURE NO. In this particular example, the experimenter has chosen to run just 10 creatures, and it so happens the 3rd one is on the screen at the moment.

Figure 6-2B shows the screen format after the experiment is done. In this case, the data represents the compiled results of running 10 cycles or creatures in succession. The numbers show the average scores after 10, 20, 30, ..., 100 contacts. This is the data that makes up the learning curve for the experiment.

FLOWCHART FOR ALPHA-I COMPILE

The flowchart for this project is shown in Fig. 6-3. The general flow of operations is quite similar to that of ALPHA-I

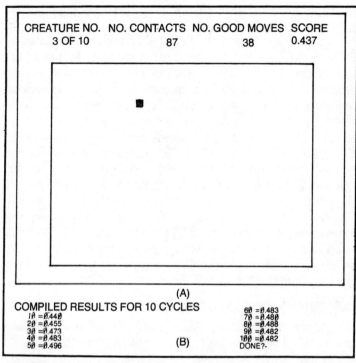

CREATURE NO.   NO. CONTACTS   NO. GOOD MOVES   SCORE
    3 OF 10   87   38   0.437

(A)

COMPILED RESULTS FOR 10 CYCLES
1∅ =∅.44∅   6∅ =∅.483
2∅ =∅.455   7∅ =∅.48∅
3∅ =∅.473   8∅ =∅.488
4∅ =∅.483   (B)   9∅ =∅.482
5∅ =∅.496   1∅∅ =∅.482
                                       DONE?-

Fig. 6-2. Screen formats for ALPHA-I COMPILE. (A) Creature presentation. (B) Compiled data.

with score (Fig. 4-2). That should come as no big surprise, considering this project is built upon the scoring program in Chapter 4. The only points of difference, as far as operations are concerned, are the addition of an S.9∅∅ COMPILE RUN, a RUN DONE conditional operation, a SERIES DONE conditional, a COMPILE SERIES display operation, and an AGAIN conditional.

S.9∅∅ COMPILE RUN is the operation responsible for dividing each creature's experience into sets of 10 contacts and accumulating the results in memory. RUN DONE signals whether or not a given creature has completed its 100-contact life cycle. If the number of contacts is less than 100, this conditional operation lets the creature continue its playing around on the screen. Otherwise, control is shifted over the SERIES DONE conditional.

SERIES DONE senses whether or not the experiment has run the designated number of creatures. If, for example, the experimenter designates a series of 100 creatures, this conditional is not satisfied until the 100th creature lives out its 100-contact cycle.

Suppose a creature has completed its run, as determined by RUN DONE, but it is *not* the final creature to be run in the experiment. SERIES DONE thus generates an N result, and the program sets up a whole new creature format by going back up to the DRAW FIELD and INITIALIZE RUN operations.

When the final creature completes its 100-contact life, the SERIES DONE conditional is satisfied and control goes to the COMPILE SERIES operation. This operation is the one that does the mathematical work and formatting necessary for generating the scores. After printing out the compiled results of the entire experiment, control goes to the AGAIN conditional. This gives the experimenter a chance to start a whole new experiment by striking the ENTER key. The most important effect of this operation, however, is to hold the compiled results on the screen for an indefinitely long period of time — until you get back from that fishing trip, for instance.

SUBROUTINES FOR ALPHA-I COMPILE

As mentioned earlier, ALPHA-I COMPILE is built around the ALPHA-I with scoring programs described in Chapter 4. So if you have saved that program and its subroutines on cassette tape, you won't have much trouble putting together ALPHA-I COMPILE. The subroutines that carry over from ALPHA-I with scoring are:

    S.50 DRAW FIELD (FIELD-1)
    S.600 INITIALIZE RUN (INITIAL-2)
    S.650 FETCH NEW MOVE (FETCH NEW-1)
    S.725 CONTACT (CON SENSE-1)

Subroutines carrying over with slight modifications are:

    S.500 HEADING
    S.800 UPDATE SCORE

The only new subroutines are:

    S.625 INITIALIZE SERIES
    S.900 COMPILE RUN
    S.875 COMPILE SERIES

Modifications for S.500 HEADING

The only operational difference is the addition of an INPUT statement in line 515. This is where the experimenter enters the

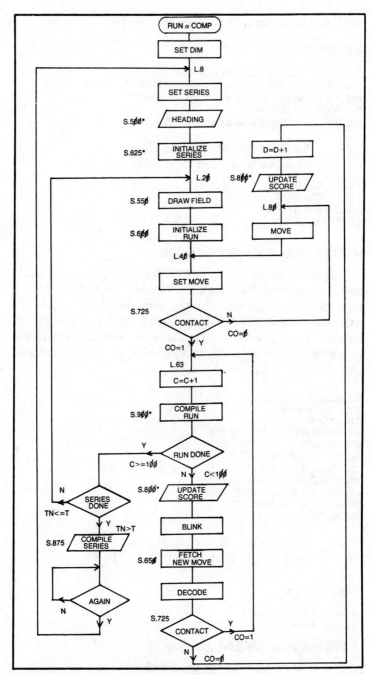

Fig. 6-3. Flowchart for ALPHA-I COMPILE.

number of creatures to be studied, variable T.

```
500 REM ** COMP HEAD-1 **
505 CLS
510 PRINTTAB (25)"RODNEY ALPHA-I":PRINT
515 PRINTTAB (15)"BASIC VERSION WITH SCORE COMPILING"
517 PRINT:INPUT"HOW MANY TEST CYCLES";T
520 FORL=ØTO1Ø:PRINT:NEXT
525 INPUT"DO ENTER TO START";S$:RETURN
```

> Variables introduced: T, L, S$
> Variables required at start: none

Test routine
```
1 CLS:GOSUB5ØØ
2 CLS:PRINTT,"OK"
3 INPUTX$:GOTO1
```

Modification for S.800 UPDATE SCORE

This subroutine has to be modified to display the creature number—see lines 845 and 857. Variable TN in line 857 is a tally of the number of creatures that have been run in the experiment, and variable T is the total number of creatures that have to be run in order to complete the experiment. You should recall that the value of T is established by the experimenter as part of the revised HEADING subroutine.

```
800 REM ** UD SCORE COMP-1 **
805 U$="#.###"
810 E=D/C
845 PRINT @ Ø,"CREATURE NO."
850 PRINT @ 15,"NO. CONTACTS"
852 PRINT @ 35,"NO. GOOD MOVES"
853 PRINT @ 55,"SCORE"
855 PRINT @ 87,C:PRINT @1Ø8,D
857 PRINT @ 64,TN" OF"T
860 PRINT @ 119,USINGU$;E
865 RETURN
```

> Variables introduced: U$, E
> Variables required at start: D, C, TN, T

Test routine

```
1 CLS:C=1:GOSUB8ØØ
2 INPUTX$:GOTO1
```

S.625 INITIALIZE SERIES Subroutine

This is a new subroutine which must be entered from scratch. Its only purpose is to initialize a portion of computer memory devoted to accumulating the statistical data for the experiment. As shown later in this chapter, the data-accumulating operation is carried out by means of a 10 × 2 array. This subroutine is responsible for initializing all parts of the array to zero.

```
625 REM ** INITIAL COMP-1 **
630 FORN=1TO1Ø:FORM=1TO2:T(N,M)=Ø:NEXT:NEXT
635 RETURN
```

> Variables introduced: M,N
> Variables required at start: none

S.900 COMPILE RUN Subroutine

This new subroutine appears rather simple at first glance, but it does a big job as far as the compiling operations are concerned. Line 9Ø5 is responsible for determining whether or not it is time to do any compiling. Specifically, it looks for contacts 10, 20, 30, ..., 100. This is done by comparing $C/1Ø$ with the integer value of $C/1Ø$. If C is *not* a multiple of 10, the result of dividing C by 10 will be some number containing significant decimal values. If C is equal to 45, for example, $C/10$ is 4.5 — the result carries a significant decimal value of 5. However, the integer value of $C/10$ is simply 4. And since the two numbers aren't equal, line 9Ø5 in this subroutine calls for returning to the main program without doing anything else to the data.

If, on the other hand, C is a multiple of 10, the process of dividing C by 10 yields an integer value that equals $INT(C/1Ø)$. Say, for example, C is equal to 50. Dividing that number by 10 gives us a 5 — an integer equal to $INT(5Ø/1Ø)$ or $INT(5)$.

Whenever C is a multiple of 10, program control goes to line 91Ø where the current values of C and D are accumulated in the program's 10 × 2 array. In this line, TF is defined as the integer value of $C/1Ø$. The value of the first element in array $T(TF,1)$ is then increased by C, after which the current value of D is added to its old value in array $T(TF,2)$.

What develops is a list of ten items (TF=1 through 10), each having two elements. The first element in each case is the accumulated value of C $(TF,1)$, and the second element in each case is the accumulated value of D $(TF,2)$.

After accumulating the values of C and D in line 91Ø, the system returns to the main program without making any further changes in the status of the experiment, assuming TF is less than 10. If TF is less than 10, the creature on the screen has not yet completed its 100-contract cycle.

If, on the other hand, TF is equal or greater than 10, the creature has completed its useful life, and it is time to start with another creature. Line 920 increments the creature number (TN) before returning operations to the main program.

In summary, 900 COMPILE RUN does three things:

1. It determines whether or not it is time to accumulate the current values of C and D.
2. Accumulates the values of C and D in a 10 × 2 array whenever it is time to do so.
3. And increments the creature count (TN) whenever one completes its 100-contact life cycle.

```
900 REM ** COMPILE-1 **
905 IFC/10<>INT(C/10)RETURN
910 TF=INT(C/10):T(TF,1)=T(TF,1)=T(TF,1)+C:T(TF,2)=T(TF,2)+D
915 IFTF<10RETURN
920 TN=TN+1:RETURN
```

> Variables introduced: TF
> Variables required at start: C, D, TN

S.875 Compile Series

This subroutine prints out the results of the experiment, formatting it as shown in Fig. 6-2B. It first prints the heading message, including the number of creatures or cycles that were run. Line 885 then reads the array of accumulated data, printing the contact decades, followed by an equal sign, and the results of dividing the accumulated D values by the accumulated C values. The result is a set of 10 scores which represent the average scores for TN creatures in steps of ten contacts. The final line in this subroutine does the AGAIN conditional operation shown on the flowchart in Fig. 6-3.

```
875 REM ** PRINT COMP-1 **
880 CLS:PRINT"COMPILED RESULTS FOR"TN-1"CYCLES":PRINT:PRINT
885 FORN=1TO10:FC=T(N,1):FD=T(N,2)
890 PRINTN*10"=";:PRINTUSINGU$;FD/FC:NEXT
895 PRINT:PRINT:INPUT"DONE";X$:RETURN
```

> Variables introduced: N, FC, FD, X$
> Variable required at start: T(N,1), T(N,2), U$, TN

THE MASTER PROGRAM FOR ALPHA-I COMPILE

The master program for ALPHA-I COMPILE is a slightly revised version of the ALPHA-I with scoring master program in Chapter 4. So, again, if you have saved the ALPHA-I scoring program on cassette tape, you can fix up this new compile version from it.

```
5 REM ** ALPHA-I COMPILE MASTER **
7 DIMT(10,2)
8 TN=1
10 GOSUB500
15 CLS
17 GOSUB625
20 GOSUB550
30 GOSUB600
40 X=X+I:Y=Y+J
50 GOSUB725
60 IFCO=0THEN80
63 C=C+1:GOSUB900
64 IFC>=100THEN90ELSEGOSUB800
65 SET(X,Y):SET(X+1,Y):):RESET(X,Y):RESET(X+1,Y):GOSUB650
70 I=RI−3:J=RJ−3GOSUB725
75 IFCO=1THEN63
78 D=D+1:GOSUB800
80 SET(X,Y):SET(X+1,Y):RESET(X,Y):RESET(X+1,Y)
85 GOTO40
90 IFTN<=TTHEN20
95 GOSUB875
100 GOTO8
```

Of course, line 5 has to be edited to reflect the fact that this program is different from the master in Chapter 4. Line 7 then follows the flowchart in Fig. 6-3 by setting the dimensions of the 10 × 2 array. (This is the array that serves as an accumulator for C and D values throughout the entire experiment). Line 8 then initializes the creature number, TN, to 1. Lines 10 and 15 are the same as before, but line 17 has to be inserted in order to initialize the values of all elements in the accumulator array to zero.

There are no changes in the master program until it is time to insert line 63. This line calls S.900 COMPILE RUN. Line 64 is then inserted to test the experiment for contact values of 100 — RUN DONE conditional operation.

Lines 90, 95, and 100 are added to the master program from Chapter 4. Line 90 takes care of the SERIES DONE conditional operation, while line 85 calls the S. 875 COMPILE SERIES routine and the AGAIN conditional step. If you wish to start another set of experiments, line 100 returns program operations back to line 8. Everything else in this master program follows the ALPHA-I with scoring program exactly. Even the line numbers are the same.

LOADING ALPHA-I WITH COMPILE

Load the entire ALPHA-I with scoring program into your system and then modify S.500, S.800, and the master program as described in the two previous sections of this chapter. Add subroutines S.625, S.900, and S.875.

The program is now ready to run. Take a shot at it by entering just one creature cycle the first time. Watch the ALPHA creature work its way through a hundred contact situations. The process might take about 10 minutes, but at the end of that time, the system should display a complete rundown of the scoring at 10-contact intervals. If you find any problems with the operation of the program, check your listing against the composite listing at the end of this chapter.

SOME NOTES ON RUNNING ALPHA-I COMPILE

Bear in mind that ALPHA-I COMPILE is a genuine scientific experiment, and its main purpose is to generate a learning curve which represents the learning ability of your ALPHA-I creature. It can be used as a standard for comparing the behavior of other creatures for a long time to come.

Number of Creatures to Test

The larger the number of creatures you test, the smoother and more reliable your results will be. If you run only 10 creatures, for example, it only takes one or two unusually good (or bad) ones to throw all the data out of kilter.

A reasonable minimum number of creatures is on the order of 100. Using that many creatures, the results of exceptionally high-scoring runs will be pretty much balanced out by an equal number of unusually low-scoring creatures.

In statiticians terms, you want a sampling population that is as large as possible. There are some practical limitations to how large the sampling population can be, and the most significant one is the amount of time required to run the experiment. On the average, it takes about 10 minutes to run one ALPHA-I creature through its 100-contact lifetime. Some take less time and others take more, but you can figure the total experimentation time using 10 minutes per critter as a ballpark figure.

So how long does it take to run this experiment for 100 creatures? At 10 minutes apiece, it figures out to something on the order of 1000 minutes, or about 16 hours! As suggested earlier in this chapter, you should start the experiment, spend the rest of the day fishing, and then get a good night's sleep. Then maybe the experiment will be over.

It's a good idea to hold off the experiment until you have a block of time available for running the machine. Interrupting the program to do some other sort of work on the machine wipes out the project, making it necessary to start over some other time.

In spite of the amount of computer time that's tied up with this experiment, you must do it for yourself sooner or later. Don't take my word that your creatures will behave exactly as shown in Fig. 6-1. That's not very scientific.

Any experiment worth its salt is reproducable by anyone who follows the instructions. Get into the habit of doublechecking and duplicating all the experiments described in this book. Later on, you are going to be seeing some rather incredible suggestions about the power of self-programming, adaptive machines, and you will be in a position to substantiate the claims only if you have proven every step for yourself.

Accounting for a Statistical Glitch

Notice in Fig. 6-1 that the ALPHA-I creatures tend to score somewhat better through the first 30 contacts than they do later on in the runs. You will also find this slight drop in performance in your own data. Does this mean the creatures really do better at the start? Does it mean they are smarter at first? No, not at all.

The slight, initial drop in the learning curve is an anomaly created by the way the creatures are initialized by the S.6ØØ INITIALIZE subroutine. The creatures always begin by moving downward and slightly to the right of the border figure. As a result of this initial motion, the creature's first contact with the border never involves a score-dropping corner contact. The chances of making a successful move after the initial contact is 14:24, and the implication is that the creature will usually score better than at any later time in the experiment. The situation tends to "bias" the data in the early going.

One way to eliminate the problem is to restructure the initialization subroutine so that the creature has just as good a chance of making a corner contact as one with a section of border that is far removed from a corner. Here's something along this line that is worth trying:

```
600 REM ** INITIAL COMP-3 **
605 GOSUB650
610 X =RI+25:Y=RJ+25:GOSUB650
612 I=RI-3:J=RJ-3
615 C=Ø:D=Ø
620 RETURN
```

Here, the initial position and direction of motion for the creature are both determined at random. It is a matter of calling S.65Ø FETCH NEW-1 for an entirely different purpose. After doing S.65Ø in line 6Ø5, the values of RI and RJ are summed with a constant value of 1Ø to establish the initial position of the creature. The second part of line 61Ø calls the S.65Ø FETCH NEW-1 subroutine again, but this time the values of RI and RJ are used for setting the initial direction of motion.

Run ALPHA-I COMPILE with this revised initialization subroutine and see if the learning curve is almost perfectly straight all the way across. If that's the case, we have successfully accounted for the slight anomaly in the curve of Fig. 6-1.

COMPOSITE PROGRAM LISTING FOR ALPHA-I COMPILE

```
5 REM ** ALPHA-I COMPILE MASTER **

7 DIMT(10,2)

8 TN=1

10 GOSUB500

15 CLS

17 GOSUB625

20 GOSUB550

30 GOSUB600

40 X=X+I:Y=Y+J

50 GOSUB725

60 IFCO=0THEN80

63 C=C+1:GOSUB900

64 IFC>=100THEN90ELSEGOSUB800

65 SET(X,Y):SET(X+1,Y):RESET(X,Y):RESET(X+1,Y):GOSUB650

70 I=RI-3:J=RJ-3:GOSUB725
```

```
75 IFCO=1THEN63
78 D=D+1:GOSUB800
80 SET(X,Y):SET(X+1,Y):RESET(X,Y):RESET(X+1,Y)
85 GOTO40
90 IFTN<=TTHEN20
95 GOSUB875
100 GOTO8
500 REM ** COMP HEAD-1 **
505 CLS
510 PRINTTAB(25)"RODNEY ALPHA-i":PRINT
515 PRINTTAB(15)"BASIC VERSION WITH SCORE COMPILING"
517 PRINT:INPUT"HOW MANY TEST CYCLES";T
520 FORL=0TO10:PRINT:NEXT
525 INPUT"DO ENTER TO START";S$:RETURN
550 REM ** FIELD-1 **
552 F1=15553:F2=15615:F3=16321:F4=16383
555 FORF=F1TOF2:POKEF,131:NEXT
560 FORF=F3TOF4:POKEF,176:NEXT
565 FORF=F1TOF4STEP64:POKEF,191:NEXT
570 FORF=F2TOF4STEP64:POKEF,191:NEXT
575 RETURN
600 REM ** INITIAL-2 **
610 X=10:Y=10:I=1:J=1
615 C=0:D=0
620 RETURN
625 REM ** INITIAL COMP-1 **
630 FORN=1TO10:FORM=1TO2:T(N,M)=0:NEXT:NEXT
635 RETURN
650 REM ** FETCH NEW-1 **
660 RI=RND(5):RJ=RND(5)
665 IFRI=3ANDRJ=3THEN660
```

```
670 RETURN
725 REM ** CON SENSE-1 **
730 FORXP=2TO2+ABS( I ):FORYP=1TO1+ABS( J )
735 IFPOINT( X+SGN( I )*XP,Y+SGN( J )*YP )=-1THEN750
740 NEXT:NEXT
745 CO=0:RETURN
750 CO=1:RETURN
800 REM ** UD SCORE COMP-1 **
805 U$="#.###"
810 E=D/C
845 PRINT@0,"CREATURE NO."
850 PRINT@15,"NO. CONTACTS"
852 PRINT@35,"NO. GOOD MOVES"
853 PRINT@55,"SCORE"
855 PRINT@87,C:PRINT@108,D
857 PRINT@64,TN" OF"T
860 PRINT@119,USINGU$;E
865 RETURN
875 REM ** PRINT COMP-1 **
880 CLS:PRINT"COMPILED RESULTS FOR"TN-1"CYCLES":PRINT
    :PRINT
885 FORN=1TO10:FC=T( N,1 ):FD=T( N,2 )
890 PRINTN*10"="::PRINTUSINGU$;FD/FC:NEXT
895 PRINT:PRINT:INPUT"DONE";X$:RETURN
900 REM ** COMPILE-1 **
905 IFC/10<>INT( C/10 )RETURN
910 TF=INT( C/10 ):T( TF,1 )=T( TF,1 )+C:T( TF,2 )=T( TF,2 )+D
915 IFTF<10RETURN
920 TN=TN+1:RETURN
```

A New Sensation
and Alpha-II

All Alpha-Class machines make purely random responses to changes they sense in their environments. The ALPHA-I creature featured in Chapters 3 through 6 has only one sensory mechanism and one way to respond: It senses only light-colored segments on the screen and responds by attempting to get away from them.

Alpha-Class adaptive machines need not be limited to single sensory and response mechanisms. There can be any number of sense/response mechanisms, and there are no limits on how simple or complex those mechanisms can be. The defining feature of an Alpha-Class machine is that it makes random responses or, in a few cases, pre-programmed responses.

The heirarchy of Alpha-Class machines can be divided into subgroups such as Alpha-I, Alpha-II and so on. An Alpha-I version is the simplest possible Alpha-Class machine. As shown in the previous chapters, ALPHA-I senses two, mutually exclusive conditions in its environment, namely being in a condition of running into a barrier or in a condition where it is free from any barrier contact. Furthermore, ALPHA-I responds to any barrier-contact situation with a negative "flight" response. The only response mechanism is one aimed at getting the creature away from such barriers.

The antithesis of running into a barrier is being free from such a condition which is a more desirable condition. It responds by performing the response that works. In other words, an ALPHA-I that is free from negative conditions in its environment keeps on moving the same way.

An Alpha-II machine is a bit more sophisticated. It has the ability to distinguish a larger number of environmental conditions and react accordingly. The reactions are still randomly selected, however. Specifically, the ALPHA-II creature introduced in this chapter can sense the difference between two different kinds of lighted segments on the screen, lighted segments that flash and lighted segments that do not flash.

In this case, lighted segments that flash are considered positive, desirable elements of the environment—running into a flashing lighted segment is "good." Running into a non-flashing lighted segment, however, is still considered "bad," and ALPHA-II responds accordingly.

The "good" flashing light on the screen is going to be a nest where the creature presumably gathers nourishment or engages in some other form of activity vital to its survival. Of course, this is a computer simulation and as such isn't absolutely necessary. However, if ALPHA-II were a real machine, the "nest" could represent a battery charger or a similar kind of energy source.

Figure 7-1 shows the screen format for this ALPHA-II project. The border is made up of non-flashing line segments and is thus considered something to be avoided. The next rectangle near the center of the screen, on the other hand, flashes one time whenever (1) it is turned on and (2) the creature makes contact with any lighted segment on the screen. Whenever the nest is turned off, it no longer flashes and is treated as a barrier to be avoided.

Whenever the creature senses contact with the nest and the nest is turned on, the creature responds by resting at the nest until it is turned off. In a sense, this resting response is an inborn reflex, the first and one of a very few pre-programmed responses used anywhere in this book.

Having this one pre-programmed response built into an Alpha-Class machine does not really violate the purity of the adaptive machine. Alpha-Class machines do not have the capacity for learning how to respond to a positive influence of this kind, and in the real world, an inborn reflex of this kind would be absolutely essential to survival. Rest assured, higher-order machines described later in this book will have to learn to deal with the nest situation on their own, first by trail and error and then by purposeful intent.

THE BASIC ALPHA-II PROJECT

The ALPHA-II program is started just as the ALPHA-I programs are. Subroutines in the early part of the program draw the basic rectangular border and initialize the position and motion of the creature. A new subroutine draws the nest figure in the middle of the border area and initially sets it to an ON condition.

NO. FEEDS	NO. CONTACTS	NO. GOOD MOVES	SCORE
2	32	16	Ø.5Ø

CREATURE

ON

NEST

Fig. 7-1. Screen format for BASIC ALPHA-II.

The creature bounces around within the confines of the border figure, accumulating data for the total number of contacts, number of good moves, and an overall success score. As long as the nest is turned on, however, you will notice that the nest flashes off and on each time the creature encounters the border figure. The border figure does not flash, however, and that is the clue the creature uses to distinguish a border contact from a contact with the nest.

The nest remains turned on until the creature makes contact with it. This might take only two or three moves, or maybe twenty or more. At any rate, a contact with the nest figure also causes the nest to flash, and since the creature is built to sense flashing-light activity in its path, it knows it has made contact with the nest figure.

The creature then remains in the nest area until the nest is switched off. The program includes a timing routine which allows the nest to remain turned on several seconds after the creature makes contact with it. At the end of that time, the program turns off the nest, and it then appears as an obstacle to be avoided. The creature responds by making random motions aimed at getting away from the nest.

Shortly after the creature moves away from the nest figure, the program turns on the nest once again and the little creature bounces around within the border until it makes contact with the nest again. The nest remains on for several seconds and then goes off again. The creature moves away and the cycle starts all over again.

The NO. FEEDS label shown in Fig. 7-1 keeps track of the number of times the creature makes contact with the nest while it is turned on. In essence, it is a measure of how successful the creature is at finding the nest.

The whole project can run automatically, but for experimental purposes, it is possible to override the automatic feature of the nest and switch it on and off manually from the keyboard. The nest can be turned off at any time by striking the X key and turned on at any time by striking the F key. This manual feature gives you an opportunity to interact with the creature on the screen.

While running these ALPHA-II experiments, bear in mind that they are simulations of real situations. The border figure represents qualities of the environment which could threaten the existence of the robot—walls, cliffs, oceans, or enemy fortresses. The nest figure has a dual nature. It can be a good thing or a bad thing, depending upon its condition at the moment. In a real robot environment, the nest could represent a small source of energy which is easily depleted. As long as there is energy available (the nest is turned on), it is a "good" thing, but when the nest is turned off, the energy is gone and spending time dealing with it is a waste of valuable energy and time, energy and time that would be better spent looking for a "live" nest.

FLOWCHART FOR BASIC ALPHA-II

The BASIC ALPHA-II program is built around the ALPHA-I with scoring program described in Chapter 4. This is entirely consistent with the notion you are working with an evolutionary process one that develops only by building upon earlier, more primitive systems.

As you look over the rather extensive flowchart in Fig. 7-2, you should be able to see many processes that have become a familiar part of your earlier work with ALPHA-I programs. In fact, the flowchart is identical to the one for ALPHA-I with scoring (Fig. 4-2), until you notice the Y response from the first CONTACT conditional. This first CONTACT subroutine (the one directly following SET MOVE) informs the creature it has encountered something. The next step is to determine whether that "something" is a switched-on nest, (a negative barrier) the border figure, or a switched-off nest.

So after sensing the contact, the next conditional, NEST ON, determines whether or not the nest is turned on. If indeed the nest is turned on, the NEST ON conditional is satisfied and the nest responds by being blanked from the screen. Seen the BLANK NEST operation.

After the nest figure is blanked off by S. 990 BLANK NEST, the CONTACT subroutine is called again. The point of this is to find out whether or not the path in front of the creature has become clear. If the contact has been with a turned-on nest, this second running of the S.725 CONTACT subroutine will show a clear path in front of the creature. On the other hand, if the contact has been with the border figure, the creature's path will remain blocked by a non-flashing light.

So the second S.725 CONTACT conditional, the one following BLANK NEST, is the operation that distinguishes the turned-on nest figure from the border figure.

For the sake of this discussion, suppose the creature finds it has made contact with the turned-on nest. The N path from the second contact conditional is thus in force, and the system responds by replacing the nest figure on the screen.

The flashing effect of the nest is achieved by S.990 BLANK NEST and S.580 DRAW NEST. The first of these two subroutines blanks the nest figure from the screen, and the second restores the figure. This all happens so rapidly that you get the impression the nest figure simply blinked off for some fraction of a second.

Inserting an S.725 CONTACT subroutine between these flashing steps merely serves as a convenient and simple way to sense the difference between the nest and border figures. If, after making contact with something on the screen, the creature notes that the "something" disappears momentarily (no contact), there is good reason to believe the contact is with the turned-on nest.

Any time the nest is redrawn, it must be switched on again. This is due to a peculiarity of the program, and it is not really a vital part of the philosophy at work in the ALPHA-II scheme.

So after the nest is blinked off and on, and the creature notes that it has indeed contacted the nest, the NO. FEEDS score is incremented by the COUNT FEEDS block. The creature is then allowed to "feed" at the nest for a period of time determined by BLINK/FEED TIME AND FEED DONE. While the creature is "feeding," its figure blinks rapidly on the screen, expressing a feeling of happy excitement.

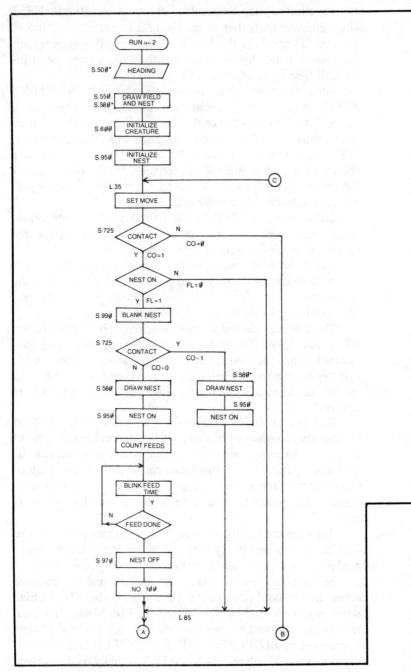

Fig. 7-2. Flowchart for BASIC ALPHA-II.

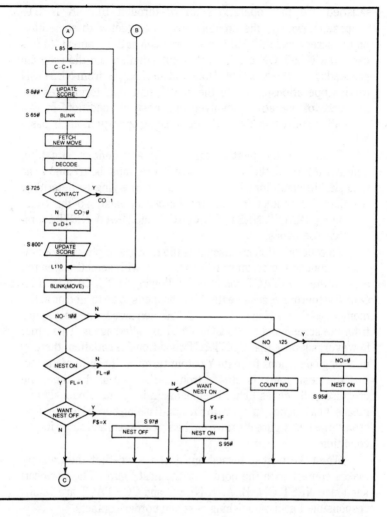

When FEED TIME is done, the nest is switched off. The nest figure is not blanked from the screen, but rather it is defined as turned off. You can note this condition by the fact that the label within the nest figure changes from ON to OFF. All this is done by the subroutine S.970 NEST OFF.

The NO variable is then initialized to a value of 100, and the flowchart enters a whole new phase of operation. The purpose of the NO variable will be described a bit later in this discussion.

Now, back up to the NEST ON conditional operation. The previous part of the discussion assumed this conditional is satisfied and that operations follow through BLANK NEST. Suppose, however, the creature makes contact with something on the screen and the nest is *not*; turned on at the moment. In this case, the NEST ON conditional is not satisfied and the program proceeds from the N output. Note that an N output from NEST ON carries operations down to the same point that terminates the sequence of operations involved in a nest-on condition. In this case, all the nest-blanking and other feeding operations are bypassed.

Whenever the nest is turned off, any contact situation, whether it be with the border or nest figure, must be treated as an undesirable condition. There is no point in trying to determine whether the contact is with the border or nest, so the time-consuming BLANK NEST, CONTACT, and DRAW NEST operations can be bypassed.

To conclude this discussion of the first major phase of operations, suppose the creature makes contact with something on the screen (the CONTACT conditional following SET MOVE is satisfied). Furthermore, assume the nest happens to be turned on at the moment (NEST ON is satisfied). S.990 then erases the nest figure from the screen and S.725 CONTACT is called again. This time, however, suppose that CONTACT conditional is satisfied, thereby directing operations from its Y output terminal. In spite of the fact that the nest figure is erased from the screen at the time, the creature still senses lighted segments in its path. Obviously this means it has made contact with the non-flashing border figure. As a result, the nest figure is drawn on the screen again and set to its ON condition.

When the nest is turned on, it blinks whether the creature makes contact with the border or the nest figure. The flowchart sequence NEST ON, BLANK NEST, and CONTACT are wholly responsible for distinguishing nest and border contacts.

The second major phase of the operation picks up at a familiar C=C+1 block. The operations following C=C+1 is the conventional Alpha-Class routine for selecting random responses and updating the score figures. If you have been doing your homework all along, this second phase, between C=C+1 and BLINK, requires no further explanation.

The final phase of the BASIC ALPHA-II program involves a number of conditional operations and housekeeping chores. Most of the operations in this final phase are devoted to the feature that allows you to turn the nest off and on manually.

Assume for the moment that the conditional operation labeled NO<=1ØØ is satisfied (Y output), the system encounters another conditional labeled NEST ON. If the nest is turned on at the moment, the next question is whether or not you want it turned off. If so, the third phase of the project is ended after doing an S.97Ø NEST OFF subroutine. Otherwise, the phase is concluded with the nest remaining on.

If, on the other hand, the NEST ON conditional is not satisfied, the implication is that the nest must be turned off. This situation brings up the question WANT NEST ON? If so, the phase ends only after doing S.95Ø NEST ON. Otherwise, the system ends this phase with the nest remaining off.

Next, consider the NO variable, the one initialized at 100 after the creature is through feeding at the nest. The overall purpose of this variable is to turn on the nest automatically after the creature has a chance to leave the nest.

The third phase of the program begins with the conditional NO <ØØ. In essence, the operation is asking, "Has the creature just completed a feeding operation at the nest?" If so, NO has to be greater than or equal to 100, because it is initialized at 100 at the conclusion of a feeding operation.

So immediately after doing a feeding operation, NO is equal to 100 and the conditional NO1<1ØØ is not satisfied. This brings up yet another conditional related to the value of NO, NO>125. If NO is less than 120 (but greater than 100), the system does an operation labeled COUNT ON. This simply an NO=NO+1 statement which executed with every move the creature makes.

Making a rather long story short, the NO counter begins running the moment the creature ends a feeding operation and continues running through 25 moves (the interval between NO=100 and NO=125. The moment NO reaches 125, it is reset to zero by NO=Ø and the rest is turned on again by S.95Ø NEST ON.

All of this gives the creature ample time to find its moves for getting away from the nest before the nest is turned on again. Unfortunately this is one of those peculiar programming situations where the basic concept is simple, but the implementation is less straightforward.

SUBROUTINES FOR BASIC ALPHA-II

Several subroutines carry over unchanged from the ALPHA-I with scoring programs in Chapter 4. These include:

```
S.550 DRAW FIELD (FIELD-1)
S.600 INITIALIZE CREATURE(INITIAL-2)
S.725 CONTACT (CON SENSE-1)
S.650 FETCH NEW MOVE (FETCH NEW -1)
```

Then, too, some of the ALPHA-I subroutines carry over with minor modifications:

```
S.500 HEADING
S.800 UPDATE SCORE
```

The entirely new subroutines are:

```
S.580 NEST FIG-1
S.950 NEST ON-1
S.970 NEST OFF-1
S.990 NEST BLANK
```

Modifications for S.500 Heading

The HEADING subroutine is merely revised to reflect the fact that the current program is an ALPHA-II NEST VERSION.

```
500 REM ** ALPHA-II HEAD-1**
505 CLS
510 PRINTTAB (25)) "RODNEY ALPHA-II":PRINT
515 PRINTTAB (25) "NEST VERSION"
520 FORL=ØTO1Ø:PRINT:NEXT
525 INPUT "DO ENTER TO START"; S$:RETURN
```

> Variables introduced: L, S$
>
> Variables required to start: none

The operating format in this case is identical to the now familiar ALPHA-I with scoring program.

Modifications for S.800 Update Score

The changes in this common subroutine involve adding data relating to the number of times the creature makes contact with the nest when it is turned on. It introduces variable NF, the number of feedings.

```
800 REM ** UD NEST-1**
805 U$="#.###"
810 E=D/C
845 PRINT @ Ø, "NO. FEEDS"
850 PRINT @ 15, "NO. CONTACTS"
852 PRINT @ 35, "NO. GOOD MOVES"
853 PRINT @ $55, "SCORE"
855 PRINT@ 87, C:PRINT 1Ø8, D
857 PRINT @ 64,NF
860 PRINT @  119, USINGU$;E
865 RETURN
```

> Variables introduced E, U$
>
> Variables required at start: D,C, NF

Simply change the REM statement in line 8ØØ and add the statements in lines 845 and 857.

S.580 DRAW NEST Subroutine

This new subroutine is a rather straightforward application of some POKE statements. The objective is to draw the nest rectangle near the middle of the border area.

```
580 REM ** NEST FIG-1 **
582 FA=15900:FORF=FATO FA+7:POKEF,176:NEXT
585 FORF=FA+64TOFA+71:POKEF,191:NEXT
587 FORF=FA+128TOFA+135:POKEF,131:NEXT
590 RETURN
```

Variables introduced: FA, F

Variables required to start: none

You can check the operation of the subroutine by running the following test routine:

```
1 CLS:GOSUB550
2 GOSUB580
3 INPUTX$:GOTO1
```

This assumes, of course, that S.550 DRAW FIELD is already resident in the program memory. You should see both the border figure and nest drawn on the screen, and you can repeat the drawing process any number of times by striking the ENTER key.

S.950 Nest On andS.970 NEST OFF Subroutines

These two subroutines are responsible for turning the nest on and off. These are *not* the operations responsible for actually blanking and restoring the nest figure. Rather, they alter the feeding status.

Whenever S.950 NEST ON is run, line 955 uses the computer's POKE feature to position the characters ON near the middle of the nest figure. This is for the convenience and enlightenment of the experimenter. The real NEST ON task is that of setting the nest flag status to 1. Whenever the nest status flag, variable FL, is equal to 1, the nest is turned on and ready to service the needs of the creature.

S.970 NEST OFF sets the feed status flag, FL, to zero (line 980) and prints the message OFF in the nest figure.

```
950 REM ** NEST ON-1 **
955 POKEFA+66C 79:POKEFA+67, 78:POKEFA+68, 128
960 FL=1:RETURN
970 REM ** NEST OFF-1 **
975 POKEFA+66, 79:POKEFA+67, 70:POKEFA+68, 70
980 FL=0:RETURN
```

Variables introduced: FL

Variables required at start: FA

If you'd like to check out these subroutines, try this test program:

```
1 CLS:GOSUB550
2 GOSUB580
3 INPUTX$
4 IFFL=0THEN7
5 GOSUB970
6 GOTO1
7 GOSUB950
8 GOTO1
```

This program first draws the border and basic nest figures, then allows you to switch the nest ON and OFF alternately by striking the ENTER key. Be sure to delete this test program before attempting to run the main program later on.

S. 9900 BLANK NEST Subroutine

This is the subroutine that actually blanks the nest figure from the screen whenever the creature is attempting to sense its presence. Like its DRAW NEST counterpart, it uses POKE graphics. In this case, however, POKE inserts spaces that effectively remove the image from the screen.

```
990 REM ** NEST BLANK **
992 FORF=FATOFA+7:POKEF, 128:NEXT
994 FORF=FA+64TOFA+71:POKEF, 128:NEXT
996 FORF=FA+128TOFA+135:POKEF,128:NEXT
998 RETURN
```

> Variables introduced: none
> Variables required at start: FA

THE BASIC ALPHA-II MASTER PROGRAM

Although the ALPHA-II master program is built upon the ALPHA-I with scoring program from Chapter 4, the alterations are so extensive that it is necessary to rewrite the whole thing from scratch. Fitting in a new line here and there, and altering a few other lines simply creates an unwieldy mess.

The program follows the flowchart in Fig. 7-2 rather closely. In fact, if you've had some trouble following the rationale of the flowchart, you might get some valuable insight by comparing it with this master program listing.

```
5 REM ** ALPHA-II NEST MASTER **
10 GOSUB500
15 CLS:GOSUB550
20 GOSUB580
25 GOSUB600
30 NO=0:GOSUB950
35 X=X+I:Y=Y+ J:GOSUB725
40 IFCO=0THEN110
```

```
45 IFFL=ØTHEN85
5Ø GOSUB99Ø
55 GOSUB725
6Ø IFCO=1THEN82
65 GOSUB58Ø:GOSUB95Ø
7Ø NF=NF+1FORT= Ø TO25:SET(X,Y):SET (X+1,Y):RESET (X,Y):RE-
   SET(X+1,Y):NEXT
75 GOSUB97Ø
8Ø NO=1ØØ:GOTO85
82 GOSUB58Ø:GOSUB95Ø
85 C=C+1:GOSUB8ØØ
9Ø SET(X,Y):SET (X+1,Y):RESET (X,Y):RESET(X+1,Y):GOSUB65Ø
95 I=RI-3:J=RJ-3:GOSUB725
1ØØ IFCO=1THEN85
1Ø5 D=D+1:GOSUB8ØØ
11Ø SET(X,Y):SET(X+1,Y):RESET(X,Y):RESET(X+1,Y)
115 IFNO<1ØØTHEN135
12Ø IFNO>=125THEN13Ø
125 NO=NO+1:GOTO35
13Ø NO=Ø:GOSUB95Ø
135 IFFL=1THEN145
14Ø F$=INKEY$:IFF$="F"GOSUB95Ø
142 GOTO35
145 F$=INKEY$:IFF$="X"GOSUB97Ø
15Ø GOTO35
```

The only features of this program that aren't specifically described on the corresponding flowchart are the applications of the INKEY\$ functions in lines 14Ø and 145. These are the program operations that allow you to control the status of the nest by striking either the F or X key on the keyboard.

Lines 14Ø and 145 are directly related to WANT NEST ON and WANT NEST OFF. If the nest happens to be off, and you want it on, striking the F key will do the job as soon as program operations reach the WANT NEST ON conditional operation. Likewise, you can turn off the nest by striking the X key, and when program operations reach the WANT NEST OFF conditional, it will indeed execute the NEST OFF subroutine to do the job for you.

If you use neither of these keys and their associated INKEY\$ functions, the nest operations will remain wholly automatic, turning off the nest after the creature has fed itself for a moment and then turning the nest on again when the creature has moved away.

LOADING BASIC ALPHA-II

As mentioned several times through this chapter, the program ought to be built upon the ALPHA-I with scoring program and subroutines described in Chapter 4. You can save yourself a lot of time by loading that older program before you start making the revisions and additions required for ALPHA-II.

Be sure to delete the entire ALPHA-I master program before entering its ALPHA-II counterpart. Otherwise you are going to end up with some unwanted program statements in the listing.

WHAT YOU SHOULD EXPECT FROM THE PROGRAM

With the program fully loaded as described, start it by entering the RUN command. You should be greeted with the heading statement, and it should remain on the screen until you strike the ENTER key.

After that, the border figure should be drawn on the screen, followed by the nest figure with ON written in the middle of it. Then the creature should appear near the upper left-hand corner of the border area, moving downward and toward the right. It will not make contact with the nest figure on this first set of moves. Once the creature contacts the bottom of the screen, the scoring messages should appear across the top of the screen, and the whole thing is fully underway.

Assuming the nest is still displaying its ON status, each contact the creature makes with the border figure ought to cause the nest figure to disappear for a moment, but only long enough for the creature to sense it has hit the border and not the nest itself. Remember it is this flashing effect of the nest that lets the creature sense that it is different from the the border figure.

Eventually the creature will make contact with the nest. You will see the creature standing still at the nest, but flashing rather rapidly. Why not flash like that, he's quite happy.

The creature's moment of bliss is soon over, however, and you will see the nest status change from ON to OFF. Now the nest is something to be avoided, and the creature will respond with random motions intended to get it away from the situation.

Sooner or later the creature finds a way to get away from the nest (sometimes leaving a little bit out of it). Shortly after leaving the nest, you will see the nest status returning from OFF to ON. In the meantime, the scoring figures will show an additional nest contact and some changes in the other scores as well. The program ought to run automatically in this fashion for an indefinitely long period of time.

If you want to participate in the activity, you can always override the automatic nest-status feature by striking the F and

X keys. Unless the creature is busy trying to get out of a contact situation, depressing the F key almost immediately sets the nest status to ON. And by the same token, striking the X key sets the nest status OFF. Returning the system to its automatic function is a simple matter of striking the F key to turn on the nest.

ALPHA-II NEST WITH COMPILE

For the sake of future reference and sound scientific procedure, you ought to compile a learning curve for this nest-responsive ALPHA critter. The general operating procedures and rationale for compiling this sort of statistical information have already been described in Chapter 6. The only significant difference here is that the data includes information regarding nest contacts as well as contacts with the border figure.

The compiling program featured here runs much more slowly, partly because the program itself is a bit slower than ALPHA-I and partly because the critter spends some extra amount of time feeding at the next figure. So you should allow at least one full day to run a hundred different creatures—all automatically, of course.

Figure 7-3 shows the display as it appears during the compiling runs. CREAT NO. shows you which creature is going

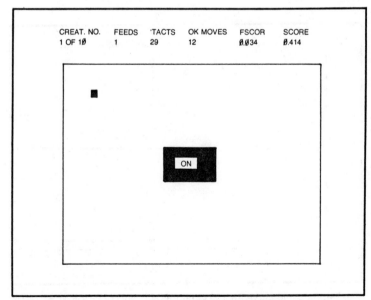

Fig. 7-3. Screen format for ALPHA-II NEST WITH COMPILE.

through its paces at the moment and the total number of creatures you have specified for the experiment.

FEEDS indicates the number of nest feedings the on-line creature has made, while 'TACTS shows the total number of contacts, both with the nest and border figure. (The number of contacts with the border figure can be calculated by subtracting FEEDS from 'TACTS).

OK MOVES shows the number of good moves the creature has made to the moment. FSCOR is the percentage of contacts that have been with the nest (FEEDS/'TACTS). While SCORE is the overall success score (OK MOVES/'TACTS.)

Figure 7-4 shows the learning curve and data display that resulted from running a hundred such creatures. Doing this yourself is a simple matter of starting the program and responding to the message, HOW MANY TEST CYCLES?, by entering 100 and then taking a hike for a day while the computer does the rest of the work.

The compiled results of the experiment shows DECADE, SCORE, and FEED. DECADE indicates which group of ten contacts (1 indicates contacts 1 through 10, 2 indicates contacts 11 through 20, etc.). SCORE is the compiled score and FEED is the compiled percentage of contacts resulting in a feeding operation.

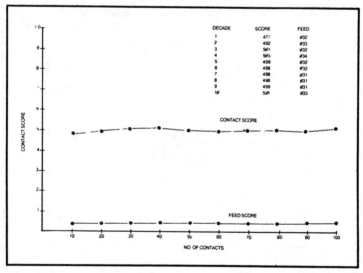

Fig. 7-4. Data and curve for 100 creatures run with ALPHA-II NEST WITH COMPILE.

The curve in Fig. 7-4, representing this data, shows a fairly steady success score of about .49 and a feed contact score of about .031.

Notice that the CONTACT SCORE does not slope downward from .54 during the first ten contacts as it does in the ALPHA learning curve back in Fig. 6-1. This flatter curve during the initial contacts is a reflection of fact that the glitch described in Chapter 6 has been ironed out. You will find the necessary correction in Subroutine 600 INITIAL COMP-3.

LOADING ALPHA-II NEST COMPILE

The following program uses many of the subroutines and the master program from BASIC ALPHA-II described in the first part of this chapter. If you have loaded BASIC ALPHA-II onto cassette tape, loading this compile version into the system is a matter of reloading BASIC ALPHA-II into your computer and making the revisions suggested here.

The subroutines used without modification are:

S.550 DRAW FIELD (FIELD-1)
S.580 DRAW NEST (NEST FIG-1)
S.950 NEST ON (NEST ON-1)
S.725 CONTACT (CON SENSE-1)
S.990 BLANK NEST (NEST BLANK)
S.970 NEST OFF (NEST OFF-1)
S.650 FETCH NEW MOVE (FETCH NEW-1)

Subroutines to be modified are:

S.500 HEADING (ALPHA-II HEAD-2)
S.600 ITIALIZE CREATURE (INITIAL COMP-3)
S.800 UPDATE DISPLAY (UD NEST-2)

The new routines are:

S.625 INITIALIZE SERIES (INITIAL COMP-2)
S.900 COMPILE RUN (COMPILE-2)
S.875 COMPILE SERIES (PRINT COMP-2)

The subroutines to be modified are shown here. In most instances, you can make the modifications with a few simple editing operations. Especially note that the starting position and initial direction codes are selected by S.600 which uses FETCH NEW MOVE (S.650) to pick these values.

```
500 REM **ALPHA-II HEAD-2**
505 CLS
510 PRINTTAB (25) "RODNEY ALPHA-II":PRINT
```

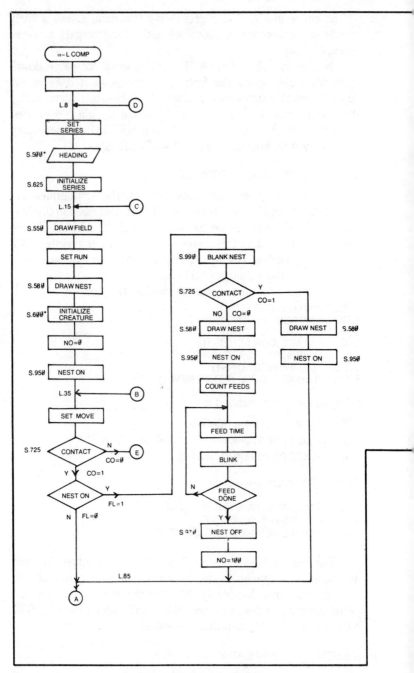

Fig. 7-5. Flowchart for ALPHA-II NEST WITH COMPILE.

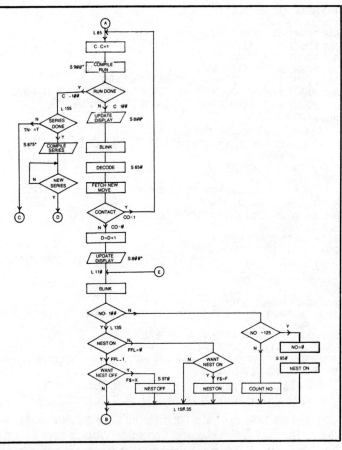

```
515 PRINTTAB (15)"NEST VERSION WITH COMPILE"
517 PRINT:INPUT"HOW MANY TEST CYCLES";T
520 FORL=ØTO8:PRINT:NEXT
525 INPUT"DO ENTER TO START";S$:RETURN
```
Variables introduced: T, L, S$
Variables required at start: none

Test routine

```
  1 GOSUB5ØØ
  2 INPUTX$:PRINTT:INPUTX$:GOTO1
600 REM ** INITIAL COMP −3 **
605 GOSUB 65Ø
610 X=RI+25:Y=RJ+25:GOSUB65Ø
612 I= RI−3:J=RJ−3
615 C=Ø:D=
620 RETURN
```
Variables required at start: None
Subroutine required to run: 65Ø FETCH NEW-1

```
800 REM**UD NEST-2 **
805 U$="#.###"
845 PRINT@ Ø, "CREAT. NO"
850 PRINT@15, "FEEDS"
852 PRINT@ 25, "'TACTS"
853 PRINT@ 35, "OK MOVES"
854 PRINT@ 47, "FSCOR":PRINT@ 57, "SCORE"
855 PRINT @ 64, TN" OF"T:PRINT @79,NF:PRINT@89,C:PRINT@ 99,D:
    PRINT @ 11 1,USINGU$;NF/CPRINT@ 121,USINGU$;D/C
865 RETURN
```
Variables introduced: U$
Variables required at start: TN, T, NF, C, D

Test routine
```
1 CLS:C=1:GOSUB8ØØ
2 INPUTX$ GOTO1
```
The new subroutines to be entered from scratch are:
```
625 REM**INITIAL COMP-2**
630 FORN=1TO1Ø:FORM=1TO3:T(N,M)= Ø:NEXT:NEXT
635 RETURN
900 REM ** COMPILE-2 **
905 IFC/1Ø<>INT(C/1Ø)RETURN
910 TF=INT(C/1Ø):T(TF,1)=T(TF,1)+C:T(TF,2)=T(TF,2)+D:T(TF,3)=T(TF,3)
    +NF
915 IFTF<1ØRETURN
920 TN=TN+:RETURN
875 REM**PRINT COMP-2**
880 CLS:PRINT"COMPILED RESULTS FOR "TN-1""CYCLES":PRINT
    :PRINT
883 PRINT"DECADE", "SCORE", "FEED
885 FORN=1TØ1Ø:FC=T(N,1):FD=T(N,2):FE=T(N,3)
890 PRINTN:PRINTUSINGU$;FD/FC:PRINTTAB(32)USINGU$;FE/FC:NEXT
895 INPUT"DONE";X$:RETURN
```
Variables introduced: N, FC, FD, FE, X$
Variables required at start: T(N,1), T(N,2), T(N,3) TN

The main controlling program is built around the one shown in
the earlier part of this chapter. Carefully compare your listing on
lines 5 through 150 with the version shown here. Insert or modify
the program as required. It really shouldn't take you very long to
get the job done.
```
5 REM ** ALPHA-II COMPILED NEST MASTER **
7 DIMT (1Ø,3)
8TN=1
10 GOSUB5ØØ
12 GOSUB625
15 CLS:GOSUB55Ø
17 NF=Ø
20 GOSUB58Ø
25 GOSUB6ØØ
30 NO=Ø GOSUB95Ø
35 X=X+I:Y=Y+J:GOSUB725
40 IFCO=ØTHEN11Ø
45 IFFL= Ø THEN85
50 GOSUB99Ø
55 GOSUB725
60 IFCO =1THEN 82
65 GOSUB58Ø:GOSUB95Ø
```

110

```
70 NF=NF+1:FORT=ØTO25:SET(X,Y):SET(X+1,Y):RESET(X,Y):RESET(X
   +1,Y):NEXT
75 GOSUB97Ø
80 NO=1ØØ:GOTO85
82 GOSUB58Ø:GOSUB95Ø
85 C=C+1:GOSUB9ØØ
87 IFC>=1ØØTHEN155ELSEGOSUB8ØØ
90 SET(X,Y):SET(X+1Y):RESET(X,Y):RESET(X+1Y):GOSUB65Ø
95 I=RI−3:J=RJ−3:GOSUB725
100 IFCO=1THEN85
105 D=Ø+1:GOSUB 8ØØ
110 SET(X,Y):SET(X+1Y):RESET(X,Y):RESET(X+1Y)
115 IFNO<1ØØTHEN 135
120 IFNO>125 THEN 13Ø
125 NO=NO+1:GOTO35
130 NO=Ø:GOSUB95Ø
135 IFFL=1THEN145
140 F$=INKEY$:IFF$="F"GOSUB95Ø
142 GOTO35
145 F$=INKEY$:IFF$="X"GOSUB97Ø
150 GOTO35
155 IFTN=THEN15
160 GOSUB875
165 GOTO8
```

If you have an doubts at all about whether or not your have entered things properly, compare your composite listing with the ALPHA-II NEST COMPILE listing at the end of this chapter.

Flowchart for Alpha-II Nest Compile

The flowchart for this compiling program is shown in Fig. 7-5. It is presented here without any special commentary. If you have been doing your homework and following the discussions, you will see it is a rather straightforward combination of the flowcharts for BASIC ALPHA-II (Fig. 7-2) and the flowchart for ALPHA-I COMPILE in Fig. 6-3.

Some Comments On Running Alpha-II Nest Compile

Once the program is initiated, it is fully automatic through the end of the project. In fact, the final results are relevant only if you do not interfere with it—keep your fingers off the keys that let you control the ON/OFF state of the nest manually. Let the thing run on automatic at all times. Otherwise you'll pay for your fun with some screwy results.

COMPOSITE PROGRAM LISTINGS

Basic Alpha-II Program Listing

```
5 REM ** ALPHA-ii NEST MASTER **
10 GOSUB500
15 CLS:GOSUB550
20 GOSUB580
25 GOSUB600
```

```
30 NO=0:GOSUB950
35 X=X+I:Y=Y+J:GOSUB725
40 IFCO=0THEN110
45 IFFL=0THEN85
50 GOSUB990
55 GOSUB725
60 IFCO=1THEN82
65 GOSUB580:GOSUB950
70 NF=NF+1:FORFT=0TO25:SET(X,Y):SET(X+1,Y):RESET(X,Y)
   :RESET(X+1,Y):NEXT
75 GOSUB970
80 NO=100:GOTO85
82 GOSUB580:GOSUB950
85 C=C+1:GOSUB800
90 SET(X,Y):SET(X+1,Y):RESET(X,Y):RESET(X+1,Y):GOSUB650
95 I=RI-3:J=RJ-3:GOSUB725
100 IFCO=1THEN85
105 D=D+1:GOSUB800
110 SET(X,Y):SET(X+1,Y):RESET(X,Y):RESET(X+1,Y)
115 IFNO<100THEN135
120 IFNO>=125THEN130
125 NO=NO+1:GOTO35
130 NO=0:GOSUB950
135 IFFL=1THEN145
140 F$=INKEY$:IFF$="F"GOSUB950
142 GOTO35
145 F$=INKEY$:IFF$="X"GOSUB970
150 GOTO35
500 REM ** ALPHA-II HEAD-1 **
505 CLS
510 PRINTTAB(25)"RODNEY ALPHA-II":PRINT
515 PRINTTAB(25)" NEST VERSION"
520 FORL=0TO10:PRINT:NEXT
525 INPUT"DO ENTER TO START";S$:RETURN
550 REM ** FIELD-1 **
```

112

```
552 F1=15553:F2=15615:F3=16321:F4=16383
555 FORF=F1TOF2:POKEF,131:NEXT
560 FORF=F3TOF4:POKEF,176:NEXT
565 FORF=F1TOF4STEP64:POKEF,191:NEXT
570 FORF=F2TOF4STEP64:POKEF,191:NEXT
575 RETURN
580 REM ** NEST FIG-1 **
582 FA=15900:FORF=FATOFA+7:POKEF,176:NEXT
585 FORF=FA+64TOFA+71:POKEF,191:NEXT
587 FORF=FA+128TOFA+135:POKEF,131:NEXT
590 RETURN
600 REM ** INITIAL-2 **
610 X=60:Y=10:I=0:J=1
615 C=0:D=0
620 RETURN
650 REM ** FETCH NEW-1 **
660 RI=RND(5):RJ=RND(5)
665 IFRI=3ANDRJ=3THEN660
670 RETURN
725 REM ** CON SENSE-1 **
730 FORXP=2TO2+ABS(I):FORYP=1TO1+ABS(J)
735 IFPOINT(X+SGN(I)*XP,Y+SGN(J)*YP)=-1THEN750
740 NEXT:NEXT
745 CO=0:RETURN
750 CO=1:RETURN
800 REM ** UD NEST-1 **
805 U$="#.###"
810 E=D/C
845 PRINT@0,"NO. FEEDS"
850 PRINT@15,"NO. CONTACTS"
852 PRINT@35,"NO. GOOD MOVES"
853 PRINT@55,"SCORE"
```

113

```
855 PRINT@87,C:PRINT@108,D

857 PRINT@64,NF

860 PRINT@119,USINGU$;E

865 RETURN

950 REM ** NEST ON-1 **

955 POKEFA+66,79:POKEFA+67,78:POKEFA+68,128

960 FL=1:RETURN

970 REM ** NEST OFF-1 **

975 POKEFA+66,79:POKEFA+67,70:POKEFA+68,70

980 FL=0:RETURN

990 REM ** NEST BLANK **

992 FORF=FATOFA+7:POKEF,128:NEXT

994 FORF=FA+64TOFA+71:POKEF,128:NEXT

996 FORF=FA+128TOFA+135:POKEF,128:NEXT

998 RETURN
```

Alpha-II Nest Compile Composite Listings

```
 5 REM ** ALPHA-II COMPILED NEST MASTER **

 7 DIMT(10,3)

 8 TN=1

10 GOSUB500

12 GOSUB625

15 CLS:GOSUB550

17 NF=0

20 GOSUB580

25 GOSUB600

30 NO=0:GOSUB950

35 X=X+I:Y=Y+J:GOSUB725

40 IFCO=0THEN110

45 IFFL=0THEN85

50 GOSUB990

55 GOSUB725
```

114

```
60 IFCO=1THEN82
65 GOSUB580:GOSUB950
70 NF=NF+1:FORFT=0TO25:SET(X,Y):SET(X+1,Y):RESET(X,Y)
   :RESET(X+1,Y):NEXT
75 GOSUB970
80 NO=100:GOTO85
82 GOSUB580:GOSUB950
85 C=C+1:GOSUB900
87 IFC>=100THEN155ELSEGOSUB800
90 SET(X,Y):SET(X+1,Y):RESET(X,Y):RESET(X+1,Y):GOSUB650
95 I=RI-3:J=RJ-3:GOSUB725
100 IFCO=1THEN85
105 D=D+1:GOSUB800
110 SET(X,Y):SET(X+1,Y):RESET(X,Y):RESET(X+1,Y)
115 IFNO<100THEN135
120 IFNO>=125THEN130
125 NO=NO+1:GOTO35
130 NO=0:GOSUB950
135 IFFL=1THEN145
140 F$=INKEY$:IFF$="F"GOSUB950
142 GOTO35
145 F$=INKEY$:IFF$="X"GOSUB970
150 GOTO35
155 IFTN<=TTHEN15
160 GOSUB875
165 GOTO8
500 REM ** ALPHA-ii HEAD-2 **
505 CLS
510 PRINTTAB(25)"RODNEY ALPHA-iI":PRINT
515 PRINTTAB(15)" NEST VERSION WITH COMPILE"
517 PRINT:INPUT"HOW MANY TEST CYCLES";T
520 FORL=0TO8:PRINT:NEXT
```

```
525 INPUT"DO ENTER TO START";S$:RETURN
550 REM ** FIELD-1 **
552 F1=15553:F2=15615:F3=16321:F4=16383
555 FORF=F1TOF2:POKEF,131:NEXT
560 FORF=F3TOF4:POKEF,176:NEXT
565 FORF=F1TOF4STEP64:POKEF,191:NEXT
570 FORF=F2TOF4STEP64:POKEF,191:NEXT
575 RETURN
580 REM ** NEST FIG-1 **
582 FA=15900:FORF=FATOFA+7:POKEF,176:NEXT
585 FORF=FA+64TOFA+71:POKEF,191:NEXT
587 FORF=FA+128TOFA+135:POKEF,131:NEXT
590 RETURN
600 REM ** INITIAL COMP-3 **
605 GOSUB650
610 X=RI+25:Y=RJ+25:GOSUB650
612 I=RI-3:J=RJ-3
615 C=0:D=0
620 RETURN
625 REM ** INITIAL COMP-2 **
630 FORN=1TO10:FORM=1TO3:T(N,M)=0:NEXT:NEXT
635 RETURN
650 REM ** FETCH NEW-1 **
660 RI=RND(5):RJ=RND(5)
665 IFRI=3ANDRJ=3THEN660
670 RETURN
725 REM ** CON SENSE-1 **
730 FORXP=2TO2+ABS(I):FORYP=1TO1+ABS(J)
735 IFPOINT(X+SGN(I)*XP,Y+SGN(J)*YP)=-1THEN750
740 NEXT:NEXT
745 CO=0:RETURN
750 CO=1:RETURN
```

116

```
800 REM ** UD NEST-2 **

805 U$="$,###"

845 PRINT@0,"CREAT. NO."

850 PRINT@15,"FEEDS"

852 PRINT@25,"'TACTS"

853 PRINT@35,"OK MOVES"

854 PRINT@47,"FSCOR":PRINT@57,"SCORE"

855 PRINT@64,TN" OF"T:PRINT@79,NF:PRINT@89,C:PRINT@99,D:
    PRINT@111,USINGU$;NF/ C:PRINT@121,USINGU$;D/C

865 RETURN

875 REM ** PRINT COMP-2 **

880 CLS:PRINT"COMPILED RESULTS FOR"TN- 1"CYCLES":PRINT
    :PRINT
883 PRINT"DECADE","SCORE","FEED

885 FORN=1TO10:FC=T(N,1):FD=T(N,2):FE=T(N,3)

890 PRINTN,:PRINTUSINGU$;FD/FC,:PRINTTAB( 32 )USINGU$;FE/FC
    :NEXT
895 INPUT"DONE";X$:RETURN

900 REM ** COMPILE-2 **

905 IFC/10<>INT( C/10 )RETURN

910 TF=INT(C/10 ):T( TF,1 )=T( TF,1 )+C:T( TF,2 )=T( TF,2 )+D:T(TF
    ,3 )=T( TF,3 )+NF
915 IFTF<10RETURN

920 TN=TN+1:RETURN

950 REM ** NEST ON-1 **

955 POKEFA+66,79:POKEFA+67,78:POKEFA+68,128

960 FL=1:RETURN

970 REM ** NEST OFF-1 **

975 POKEFA+66,79:POKEFA+67,70:POKEFA+68,70

980 FL=0:RETURN

990 REM ** NEST BLANK **

992 FORF=FATOFA+7:POKEF,128:NEXT

994 FORF=FA+64TOFA+71:POKEF,128:NEXT

996 FORF=FA+128TOFA+135:POKEF,128:NEXT

998 RETURN
```

117

8 Gearing Up For More Alpha-II Activity

The ALPHA-II schemes described in the previous chapter are intended to show a smooth transition from Level-I to Level-II programming. The techniques are rather cumbersome, however, and they promise to get worse as one attempts to apply Level-II features to the higher-order BETA and GAMMA creatures.

So it is necessary to change gears at this point and introduce some new programs and subroutines which are better equipped to deal with multiple sensory and response mechanisms. You will find this gear-changing routine being used throughout this book. All of the BETA-I and GAMMA-I programs, for example, evolve quite naturally and easily from earlier Level-I programs. Whatever you do with a Level-I creature at one point carries over to more sophisticated levels later on. By the same token, the programs and subroutines for ALPHA-II creatures, the ones presented in this chapter, carry over to BETA-II and GAMMA-II schemes.

While the flowcharting is much the same, the actual programming techniques are quite different for Level-I and Level-II creatures. In fact, as suggested in Chapter 2, you might choose to skip over any further Level-II work until you have a chance to do *all* the Level-I projects. Nothing will be lost in either case—it's up to you.

An Alpha-II Conversion Demo Program

This program demonstrates the process for converting from standard Level-I to Level-II programming. The creature in this case works very much like an ALPHA-I BASIC critter, and you should study, load, test, and save this program. You will be using it later in this chapter, and elsewhere in this book, as the starting point for building more significant Level-II creatures.

A New Alpha-II Flowchart and Master Program

Figure 8-1 shows the master-program flowchart for the improved ALPHA-II creature. If you compare this flowchart with the ALPHA-I BASIC flowchart in Fig. 3-1, you won't find many operational differences. Even the subroutine numbers are the same. The only difference of any importance is the fact that the new flowchart does not include a separate block for SET NEXT MOVE. This particular operation, as you will see shortly, is now included in MOVE.

Eliminating SET NEXT MOVE is just one of a number of changes in the programming format that are intended to reduce the amount of required program memory and, at the same time, speed up the operations.

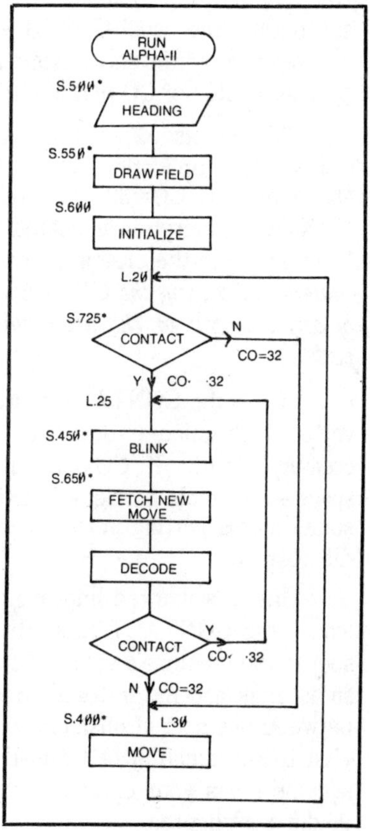

Fig. 8-1. Flowchart for ALPHA-II CONVERSION DEMO.

To get acquainted with the basic nature of the changes, study this new ALPHA-II master program:

```
10 REM ** ALPHA-II MASTER 1 **
15 CLS:GOSUB500:GOSUB550:GOSUB600
20 GOSUB725:IFCO=32THEN30
25 GOSUB450:GOSUB650:1=R1−3:60SUB725=IF00<>32 THEN 25
30 GOSUB400:GOTO20
```

It doesn't require a whole lot of BASIC programming savvy to see that this master program is much "tighter" than the corresponding ALPHA-I, BASIC-1 MASTER in Chapter 3.

Line 15 of ALPHA-II MASTER 1, for example, is a multiple-statement line which includes a CLS statement, to clear the screen initially, as well as the first three major subroutine calls. This one line clears the screen, calls the HEADING subroutine, draws the border figure, and initializes the creature.

Line 20 then calls a revised version of the CONTACT subroutine (the usual GOSUB 725), but contains the first major programming difference. Note that the CONTACT subroutine returns a value of CO that might be equal to 32 instead of 0.

The number 32, you see, is the ACSII code number for a space. The implication is that the creature sees blank spaces ahead of it if CO=32. Any other number coming out of the CONTACT operation indicates there is something other than a blank space in the creature's path. *CO, the variable that is set to some value during the CONTACT operation, indicates the nature of any numeral or character that happens to be in the creature's path.*

During the CONTACT subroutine, CO is set to some ASCII value which uniquely describes the natre of any obstacle in the creature's path. If CO=32, the system knows that a blank space—no visible barrier—exists. Otherwise, the system knows something is getting in the way and what's more, it can identify the obstacle.

This is a marked improvement over the ALPHA-I scheme where the CONTACT subroutine looks for *any* kind of lighted spot in the creature's path yet cannot tell the difference between an asterisk and the letter *Z*. Being able to sense the difference between one sort of obstacle or figure and another is absolutely vital to the function of Level-II creatures; this ASCII oriented scheme gives your creatures the potential for discerning up to 96 different figures.

Comparing this idea with the nest-sensing scheme in Chapter 7, you will find it is no longer necessary to make the nest figure blink on and off in order to distinguish it from the border figure. Simply construct the nest from some keyboard characters that are different from those used for making the border figure and the creature won't have any trouble distinguishing between the two.

Working with this new sensing mechanism calls for a knowledge of ASCII codes. You must know, for instance, that a space in ASCII is represented by decimal numeral 32 or that decimal 42 in ASCII represents an asterisk character. Appendix I in the back of this book summarizes the ASCII code in a fashion that is appropriate for these robot-intelligence projects.

Line 20 in the ALPHA-II MASTER 1 program should have more meaning to you. It says, "Do the CONTACT subroutine to pick up the ASCII value of anything in the path ahead of the creature. If that value is 32 (blank space) the road is clear and it's OK to MOVE (subroutine 400). After moving, return to line 20 to check the path again."

In this case, finding an ASCII code not equal to 32 implies that something is in the path ahead of the creature. Variable CO, in other words, is equal to something other than 32.

The program responds to ASCII contacts other than 32 by doing a BLINK, followed by FETCH NEW MOVE, DECODE, and yet another test of the path ahead. This second CONTACT subroutine is called by the final statement in line 25. Here the program is saying, "If the path ahead is still not clear (CO is equal to something other than ASCII 32), go back to the beginning of the line and pick another motion code." The system continues cycling through line 25 of the master program until it comes across a motion code, or values of I and J, that get the creature away from whatever lays in its path.

At this point in the project, the creature is searching for a clear path to get away from any obstacle, and it isn't at all concerned with the nature of the obstacle. You can be sure you will have a chance to build programs for determining the exact nature of the obstacle later in this chapter.

Subroutines For ALPHA-II Conversion Demo

The following sets of subroutines clearly demonstrate yet another important and new feature of Level-II work—the application of POKE and PEEK graphics. Look at S.400 MOVE

(MOVE II-1), for example. The purpose of this subroutine is to move the creature from one place to the next across the screen. The first statement in line 405 erases the present image of the creature by doing a POKE NP, 32. There are two important ideas inherent in this statement. The first is the use of an NP variable to indicate the creature's position on the screen. Using POKE and PEEK graphics makes it difficult to use the X-Y coordinate scheme which is applied to all Level-I programs. The faster POKE and PEEK graphics calls for specifying the creature's position as a number between 15360 and 16383, decimal versions of the memory space allocated to video display on the TRS-80 computer system.

The second idea included in the first statement in line 405 is the POKEing of ASCII 32, a space. In other words, the first statement in line 405 of subroutine 400 says, "Put a blank space (ASCII 32) in the position the creature currently holds (NP).

The second statement in line 405 sets up the creature's new position on the screen. Variables I and J play the same role as in earlier programs, acting as motion codes which set both the creature's direction and speed of motion. Instead of summing those motion-code components with their respective X and Y coordinates (which no longer exist), the system has to do the same sort of job in a POKE format. This means moving the creature to the left or right on the screen by summing the current value of NP with the value of I and moving the creature up or down by summing the current value of NP with the product of J times 64. Multiplying a POKE-position value by 64 effectively moves the POKE position one line in a vertical direction.

The second statement in line 405 sets the creature's new position on the screen using the same I and J motion-code format, but a different coordinate system. The third statement in line 405 calls for POKEing an ASCII value of 140 into the new creature position. What in the world is ASCII 140? It isn't on the chart in Appendix I. The Radio Shack TRS-80 uses ASCII-like numerals to specify a group of figure-drawing operations. See Appendix II. You can see from Appendix II that a 140 code prints a small square on the screen at position NP. This is the new way to draw the creature figure.

In summary, S.400 MOVE uses POKE graphics and ASCII codes to do three things: erase the image of the creature, set its new position on the screen, and draw the image of the creature in that new position.

Using POKE graphics and ASCII numbers to draw the creature brings up a whole new world of possibilities regarding the appearance of the creature. The creature had to appear as a white rectangle in all previous programs because its image is generated by SET graphics. Such programs do not allow you to build more elaborate creature images without having to resort to a lot of complicated and time-consuming SET routines. Now, however, you can make the creature look like any one of the figures in Appendix B and, for that matter, any of the characters specified in Appendix A as well. The only reason the creature is defined as TRS-80 ASCII in this case is so that it looks exactly like earlier models. You will have a chance to play with creatures having alternate appearances later in this chapter.

Now that you've had a chance to see how this new scheme it built around POKE graphics, the meaning of the other subroutines ought to make some sense. Subroutine 400 BLINK, for instance, simply makes the creature figure blink on and off one time. Line 455 of that subroutine says, "Erase the creature figure by POKEing a blank space in its current position on the screen and then restore the image by POKEing a 140."

```
400 REM ** MOVE II-1 **
405 POKENP,32:NP=NP+I+64*J:POKENP,14Ø:RETURN
450 REM ** BLINK II-1 **
455 POKENP,32:POKENP,14Ø:RETURN
500 REM ** HEAD II-1 **
505 CLS: PRINTTAB (25)"ALPHA-II":PRINTTAV(2Ø)"CONVERSION DE-
    MO":FORL=1TO8:PRINT:NEXT:INPUT"DO 'ENTER' TO START";
    S$:CLS:RETURN
550 REM ** FIELD II-1 **
555 FORF=1553TO15615:POKEF,131:NEXT
560 FORF=16193TO16255:POKEF,176:NEXT
565 FORF=15553TO16255STEP64:POKEF,191:NEXT
570 FORF=15615TO16255STEP64:POKEF,191:NEX  T
575 RETURN
600 REM ** INITIAL II-1 **
605 SF=Ø:NP=15968:GOSUB65Ø:I=RI-3:J=RJ-3:RETURN
650 REM ** FETCH NEW II-1**
655 RI=RND(5):RJ=RND(5):IFSF=Ø ANDRI=3ANDRJ=3THEN655
650 RETURN
725 REM ** CON SENSE II-1 **
730 FORSI=ABS(I)TO1STEP-1:FORSJ=ABS(J)TO1STEP-1
735 CO=PEEK(NP+SI*SGN(I)+64*SJ*SGN(J))
740 IFCO<>32RETURN
745 NEXTSJ,SI:RETURN
```

Subroutine 550 DRAW FIELD uses POKE graphics to draw the image of the border figure. Note that the FOR . . . TO statements use numbers within the video memory space for the

TRS-80. The POKE statements in each case generate the appropriate kinds of little rectangles.

The INITIALIZE subroutine, S. 6ØØ, sets the creature's initial position (NP) at 15968, a position near the middle of the border figure. This subroutine then calls the FETCH NEW MOVE subroutine to pick up random values for the initial direction of motion., You have seen this sort of initializing trick, used in the Levle-I X-Y coordinates, before in Chapter 6.

The SF variable, which is initialized to zero in line 605, will be used at a later time to allow the possibility of picking the creature's stop code (I=Ø,J=Ø). This is not permitted in any Level-I scheme, but accounts for a new mode of response behavior for Level-II creatures.

The exclusive use of POKE graphics accounts for quite an increase in the creature's operating speed, but the return speed comes about in new CONTACT subroutine. As in the case of ALPHA-I creature, the system searches the path ahead, using the current I and J motion code for determining the direction and distance of the search. Now, the speed of the search is increased threefold by the use of the POKE statement in line 375 of the S.725 CONTACT subroutine.

Incidentally, using this faster CONTACT subroutine brings up the intriguing possibility of building creatures that have an extended-range sensory capability. The critter can sense objects well ahead of its current position on the screen, evaluate the nature of those objects, and take any appropriate action long before a contact is *imminent*. In a manner of speaking, this would give the creature a special quality of long-range "vision."

Loading ALPHA-II Conversion Demo

Since every element of this program is new, it must be loaded into your computer from scratch—the whole thing.

NOTE: Do not test any portion of this program or run it until it is loaded in its entirety, doublechecked against the composite listing at the end of this chapter, *and recorded on cassette tape*. Converting to POKE and PEEK graphics always brings up the possibility of POKEing data into places in the computer memory that throw the whole business out of whack. A slight error in the CONTACT subroutine, for instance, can let the creature crawl down off the bottom of the screen and begin POKEing around in program memory space. Not only does this mess up the program, but it might latch up the entire computer. The only way to rectify the problem when it happens is by turning off the computer and restarting after a few seconds. Of course, this wipes out all the programming you've entered, and you lose a lot of valuable experimenting time.

So get that program onto cassette tape, as insurance against catastrophe. If the worst happens, you can load the program again from the tape and search for the error.

As added insurance against such bad things happening, you can write a fail-safe statement into the MOVE subroutine—IF NP>16383 OR NP<15360STOP.This sets bounds on the position of the creature, bringing the whole program to a halt if he begins running off the screen.

What to Expect from this Program

While this particular program does not express the truly unique sensory capabilities of an ALPHA-II creature, it does demonstrate the tremendous gain in speed. The screen format and apparent activity on the screen is identical to that of ALPHA-I BASIC. So what's the point of the program anyway? It demonstrates how it is possible to adjust the flowcharts in this book to work according to alternate programming schemes, and it sets the stage for building all Level-II creatures.

ALPHA-II WITH TRACK SENSE OPTION

Now that you have the basic building-block program for Level-II creatures, it is time to investigate some of the special properties. The program in this case is one that lets the creature leave behind a set of tracks, wherever it goes on the screen.

You worked with that sort of track-drawing program back in Chapter 5, but now you will find some of the nifty features of the new scheme coming into the forefront.

To get this one going in the shortest possible time, build it around the ALPHA-II CONVERSION DEMO program. Load the program into your computer and revise S.400 MOVE, S.500 HEADING and the master program as follows:

```
10 REM ** ALPHA-II MASTER 2 **
15 CLS:GOSUB500:GOSUB550:GOSUB600
20 GOSUB725:IFCO=320R(CO=42ANDTS=0)THEN30
25 GOSUB450:GOSUB650:I=RI-3:J=RJ-3:GOSUB725:IFNOT(CO=320R
   (CO=42ANDTS=0)THEN25
30 GOUSUB400:GOTO20
40 REM ** MOVE II-2 **
45 POKENP,42:NP=NO+I+64*J:POKENP,140: RETURN.
```

```
500 REM ** HEAD II-2 **
505 CLS:PRINTTAB(25)"ALPHA-II":PRINTTAB(20)"TRACK SENSE OP-
    TION"
510 PRINT:PRINT:PRINT"WANT CREATURE TO BE SENSITIVE TO
    TRACKS":PRINTTAB(10)"0- NO":PRINTTAB(10)"1- YES":INPUTTS
515 CLS:ONTS+1GOTO520,525
520 PRINT @ 960,"NOT SENSITIVE TO OWN TRACKS":RETURN
525 PRINT @ 960,"SENSITIVE TO OWN TRACKS":RETURN
```

Doublecheck your listing against the one at the end of this chapter and record it on cassette tape before running it. If all is going well, running the program should produce this message on the screen:

WANT CREATURE TO BE SENSITIVE TO TRACKS

    0 - NO

    1 - YES

?—

    If you respond by entering 0, the creature will leave tracks wherever it goes, but will not respond to them whenever it approaches earlier-drawn versions. Entering a 1 will make the creature treat its own tracks as a barrier, creating an effect quite similar to the PAINT INTO A CORNER GAME in Chapter 5.

    Two features of the new POKE scheme make this program important. First, you will notice that the "tracks" look different from the creature, itself. That was not the case in Chapter 5 where the tracks and the creature were both white rectangles. The second important point is that the creature is able to tell the difference between the border figure and its own tracks; if you elect to make the creature insensitive to its tracks, it responds differently to the border and tracks.

    As far as the tracks looking different, asterisks to be exact, look at the revised subroutine 400 MOVE. Instead of erasing the image of the creature by means of a PKE NP, 32 at the beginning of line 405, the system prints an asterisk in its place by doing a POKE NP, 42. The creature is then moved to its new NP position and redrawn as a square at POKE NP, 140. So every time the creature moves from one point on the screen to another, its old image is replaced with an asterisk.

    The extended logic statements in lines 20 and 25 of the master program are responsible for picking up the difference between clear space ahead (ASCII 32) and an asterisk (ASCII 42). In line 20, for instance, if the CONTACT subroutine returns with CO=32, the path ahead is clear and operations go immediately to line 30. Operations will likewise go to line 30 if CONTACT returns CO=42 (asterisk ahead), but TS is equal to zero. TS is a variable which is set to 1 or 0 during the HEADING operation. When TS=0, the creature is supposed to be insensitive to its own tracks, and the expression CO=42 AND TS=∅ means, "I'm running into an asterisk, but I've been told to ignore it." However, if the CONTACT subroutine called in line 20 returns a CO value that is *not* 32, or if CO=42, and TS=1, it means trouble is ahead. The creature senses some other figure or an asterisk to be dealt with. As a result, the system resorts to the FETCH NEW MOVE routines in line 25. So if this creature picks up anything in its path besides free space or an asterisk when TS=∅, it reacts as though it hit the border figure.

ALPHA-II WITH MULTIPLE-SENSE PRINTING

Figure 8-2 shows the screen format for this program. You will see that it consists of the usual border figure, a creature rectangle, and a family of four different obstacles. The point of the program is to show you in a very clear way that an ALPHA-II creature can really sense the differences between objects within its "playpen."

In this case, the creature treats all objects the same way—by running away from them. To let you know that the creature is actually able to sense the differences between them, the system prints an image of the last-sensed object near the upper left-hand corner of the screen. Figure 8-2 shows a percent sign in the corner. Apparently the creature has just moved away from the array of percent-sign figures near the lower left-hand corner of its playpen.

When it encounters the ∅s, a ∅ appears at the top of the screen, and when it runs into the array of asterisks, an asterisk appears there. The same idea holds for the pound signs and also the border figure.

The border figure, incidentally, is made up of TRS-80 ASCII 131 images along the top, 176 images along the bottom, and 191

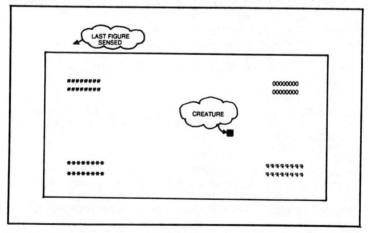

Fig. 8-2. Screen format for ALPHA-II MULTIPLE SENSE PRINTING.

images along both sides. (See Appendix II). The creature not only distinguishes the four arrays of special figures enclosed within the border figure, but distinguishes contacts with the top, bottom and sides of the border.

This program is fun to watch, but it has some powerful implications that go beyond a mere demonstration of how the creature can tell the difference between different objects. Bear in mind that one of the greatest deterrents to progress in modern robotics in the insistence upon making machines that constantly keep track of their position relative to an arbitrary frame of reference. That sort of thinking was great back in the old days when technologists were building inertial-guidance missiles and auto-pilot aircraft.

However, this little creature bouncing around on the screen in front of you can sense its position by picking up cues from objects in its playpen and from the border figure itself. An ALPHA-II is a bit too elementary to do anything significant with the sensory information of that type, but the direction for further research and development is quite clear.

The flowchart for this program is shown in Fig. 8-3. With the exception of two additional program steps, DRAW FIGS and PRINT SENSE, it is identical to the basic ALPHA-II flow-chart in Fig. 8-1. Of course, the purpose of the PRINT SENSE step is to print the image of the object the creature encounters in the upper left-hand corner of the screen. DRAW FIGS draws the figures themselves.

This program is loaded around the ALPHA-II CONVERSION DEMO program described in the first part of this chapter. First load the program from cassette tape and then make the following additions and revisions:

```
10 REM ** ALPHA-II MASTER 3 **
15 CLS:GOSUB500:GOSUB550:GOSUB675:GOSUB600
20 GOSUB725:IFCO=32THEN30
25  PRINT@ 0,CHR$(CO):GOSUB450:GOSUB650:I=RI-3:J=RJ– 3GOSUB
    725:IFC 0 < >32 THEN 25
30 GOSUB400:GOTO20
500 REM ** HEAD II-3 **
505 CLS:PRINTTAB(25)"ALPHA-II":PRINTTAB(20)"WITH SENSE PRINT"
510 FORL= 1TO8:PRINT: NEXT:INPUT"DO 'ENTER'TO  START ";S$:
    CLS:RETURN
675 REM ** FIGS II-1 **
677 P=15755
680 FORF=0 TO1:FORN=P+64*FTOP+64*F+8:POKEN,35:NEXTN,F
682 FORF=0 TO1:FORN=P+32+64*FTOP+40+64*F:POKEN,48:NEXTN,F
684 FORF=0 TO1:FORN=P+192+64*F:POKEN,42:NEXTN,F
686 FORF+0 TO1:FORN=P+224+64*FTOP+232+64*F:POKEN,37:
    NEXTN,F
690 RETURN
```

Doublecheck your modifications against the composite listing at the end of this chapter and save it on cassette tape before attempting to run it.

ALPHA-II WITH SENSE RESPONSE

The previous program illustrates the way in which an ALPHA creature can sense the quality of any obstacle in its path. In the program, however, the creature's responses are identical to every kind of object. It is possible to make a more convincing case for Level-II activity by allowing the creature to make different responses to different kinds of figures on the screen. The previous project shows how a simple ALPHA creature can have an extended sensory-input mechanism, and now you will see how it can exhibit an extended response output.

The system senses the quality of the various objects in a fashion identical to the previous program. In fact, the screen format is the same. The main difference is that this program includes a subroutine 900 which specifies the responses the creature is to make.

Actually, the notion of programming what the robot is supposed to do violates the basic principles set forth in Chapter 1. However, the point of the program is to demonstrate the creature's ability to distinguish objects of different kinds and set off an appropriate set of responses. All we are doing here is showing quite clearly that the creature is capable of originating a sequence of responses. A truly significant family of responses, in this case, can come only from a Level-III mechanism.

Loading ALPHA-II with Sense Responses

Load this program in and around ALPHA-II WITH SENSE PRINT described in the previous section. Doing so you will be able to use the following subroutines without any modifications:

```
S.400 MOVE (MOVE II-1)
S.450 BLINK (BLINK II-1)
S.550 DRAW FIELD (FIELD II-1)
S.600 INITIALIZE (INITIAL II-1)
S.650 FETCH NEW MOVE (FETCH NEW II-1)
S.675 DRAW FIGS (FIGS II-1)
S.725 CONTACT (CON SENSE II-1)
```

Next revise the master program and S.500 HEADING to:

```
10 REM ** ALPHA-II MASTER 4 **
15 CLS:GOSUB500:GOSUB675:GOSUB600
20 GOSUB725:IFCO=3THEN30
22 GOSUB900
25 GOSUB450:GOSUB650:I=RI-3:J=RJ-3:GOSUB725:IFCO<>32THEN25
30 GOSUB400:GOTO20
500 REM ** HEAD II-4 **
```

```
505 CLS:PRINTTAB(25)"ALPHA-II":PRINTTAB(18)"WITH SENSE RE-
    SPONSES"
510 FORL=1TO8:PRINT:NEXT:INPUT"DO 'ENTER'TO START";
    S$:CLS:RETURN
```

The purpose of the program is lost, however, if it isn't given something to do in response to sensing the various objects on the screen. Here is the response subroutine which has to be added:

```
900 REM ** SENSE RESP II-1 **
905 FORN=Ø TO4Ø:PRINT@N," ":NEXT:PRINT@ Ø,CHR$(CO)
910 IFCO=131ORCO=176ORCO=191PRINT@1Ø "OOF — A WALL!"
915 IFCO=35ORCO=37PRINT@1Ø,"OUCH — THAT HURTS!"
92 0 IF CO=42THEN935
925 IFCO=48THEN94Ø
930 RETURN
935 PRINT@1Ø,"OH, BOY — FUZZY THINGS":FORTD=1TO25Ø:
    GOSUB45Ø:NEXT:RETURN
940 PRINT 1Ø,"UMM, COOKIES!!":FORTD=1TO1ØØØ:NEXT:GOSUB4ØØ:
    RETURN
```

Run the program for a while and have some fun watching the creature at work and noting its comments written across the top of the border figure. Sure, it's all a bit of silliness, but there is some potential for significant work.

Some Hints About the Future of ALPHA-II

What's the future of this? What if S.9ØØ RESPONSES is rewritten to do something besides evaluate the nature of the contact situation and spell out cute little comments? What if, instead of printing comments, the system called a totally different ALPHA function, one that would establish a machine-determined response or set of responses? Indeed, it is possible to pile a number of intelligent functions into the RESPONSES subroutine, thereby giving the creature a wealth of self-determined responses to the various sensory conditions.

Unfortunately, this all adds up to Level-III work, and Level-III work is beyond the scope of this present book. It doesn't mean you are forbidden to try it, though. If you think you are prepared to handle the job, give it a shot. Make up some Alpha-Class modes of behavior to work from this sort of RE-SPONSE subroutine. Provide something more than 24 possible motion codes, and you're in the Level-III business.

GETTING TWO CRITTERS ONTO THE SCREEN AT ONE TIME

The fact that ALPHA-II creatures are able to sense a wide variety of different objects on the screen brings up the intriguing possibility of putting more than one creature onto the screen at the same time. The program featured in this section shows how this can be done.

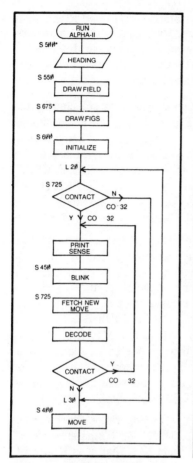

Fig. 8-3. Flowchart for ALPHA-II
MULTIPLE SENSE PRINTING.

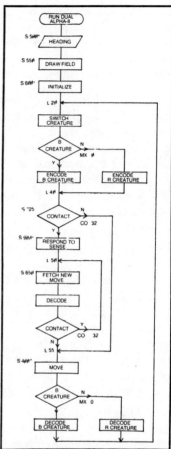

Fig. 8-4. Flowchart for ALPHA-II
WITH DUAL CREATURES.

Flowchart For ALPHA-II With Dual Creatures

Figure 8-4 is the flowchart for this two-creature program. It starts out in the usual fashion, doing things such as a HEADING, drawing the playing field, and initializing the positions and first motion codes for the two creatures.

Since the scheme is being carried out on a single computer system, the two creatures must be time multiplexed. That is to say, the two creatures share much of the program, but operate in turn. The SWITCH CREATURE block is thus an important part of this program; it switches the program treatment back and forth between the two creatures.

131

The two creatures in this instance are named R and B. In fact, they look exactly like those two alphabetical characters. The choice of names is wholly arbitrary, of course. I selected R and B because they are the first letters in the names of my two earlier robot machines, Rodney and Buster.

In any event, one of the two creatures is selected by the SWITCH CREATURE step. The B CREATURE conditional then determines which one of the two creatures is being treated at the moment. If it happens to be the B creature, the B CREATURE conditional is satisfied and ENCODE B CREATURE is called. If, on the. other hand, SWITCH CREATURE shows it is R's turn, B CREATURE is not satisfied and the system does an ENCODE R CREATURE.

The two encoding steps convert the relevant parameters for the creatures into a form that can be handled by the main program. For instance, I1, J1, and M1 represent the I, J, and NP (motion code and screen position) for the B creature. Whenever the B creature is *not* being handled by the program. those parameters are saved in memory. The ENCODE B CREATURE operation translates I1 into I, J1 into J, and M1 into NP. After that, the B creature is treated, through one entire motion cycle, as though it is the only creature on the screen. At the conclusion of B's turn, after the MOVE step, the new values of I, J, and NP are converted by DECODE B CREATURE into I1, J1 and M1.

After the DECODE B CREATURE operation, the creatures are switched at SWITCH CREATURE, and R's parameters (I2, J2 and M2) are equated with I, J and NP. Then it is R's turn to run through a motion sequence. You'll be able to see the alternating treatment on the screen—the system isn't fast enough to mask it. You will see B move, then B, then R, etc. Whenever one of them gets tangled up at the border, the other figure remains motionless until the affected creature gets out of its predicament.

In this particular program, the sensory discrimination is limited to sensing the border figure and the other creature. A message at the top left-hand corner of the screen indicates a contact between the two creatures. It shows which creature is responsible for the contact between them, and you can make a game for two players out of it if you wish. Just bet on which creature will be responsible for hitting the other more times within a given period of time.

Loading ALPHA-II With Dual Creatures

This program is loaded using ALPHA-II CONVERSION DEMO as a starting point. Load the program into your computer and modify it by making the following changes and additions:

```
10  ** ALPHA-II MASTER 5 **
15  CLS:GOSUB500:GOSUB550:GOSUB6
20  MX=MX*-1:IFMX<0THEN30
25  I=I1:J=J1:NP=M1:GOTO40
30  I=I2:J=J2:NP=M2
40  GOSUB725:IFCO=32THEN 55
45  GOSUB900
50  GOSUB650:I=RI-3:J=RJ-3:GOSUB725:IFCO<>32THEN50
55  GOSUB400
60  IFMX< 0 THEN70                 900 REM ** SENSE RESP II-2 **
65  I1=I:J1=J:M1=NP:GOTO20         905 IFCO=131ORCO=1760ORCO
70  I1=I:J2=J:M2=NP:GOTO20             =191RETURN
400 REM ** MOVE II-3 **            910 IFMX<0THEN920
405 POKENP,32:NP=NP+I+64*J         915 PRINT@0,"R HIT B":RETURN
410 IFMX<0THEN420                  920 PRINT@0,"B HIT R":RETURN
415 POKENP,82:RETURN
420 POKENP,66:RETURN
500 REM ** HEAD II-5 **
505 CLS:PRINTTAB(25)"ALPHA-II":PRINTTAB(18)"WITH  DUAL  CREA-
    TURES"
510 FORL=1TO8:PRINT:NEXT:INPUT:DO 'ENTER'TO  START";S$:
    CLS:RETURN
600 REM ** INITIAL II-2 **
605 MX=-1:SF=0:M1=15964:M2=15972
610 GOSUB650:I1=RI-3:J1=RJ-3:GOSUB650:I2=RI-3:J2=RJ-3
615 RETURN
```

Doublecheck your listing against the composite version at the end of this chapter before loading it onto cassette tape and running it.

When you run the program, you will find that the creature's move in a relatively sluggish manner. This, of course, is an effect created by the multiplexing character of the program. If you were to put three different creatures into this format, the operations would be distressingly slow. The program is presented, however, to show that it is possible to work with more than one creature at the same time. Better schemes already in the works use separate microprocessors for each of the creatures, and that allows any number of creatures to interact on the screen without any noticeable time-sharing effect. What evolves in that case is a little community of intelligent creatures. It is left to you to carry the implication to its logical conclusion.

ALPHA-II COMPOSITE PROGRAM LISTINGS

```
10 REM ** ALPHA-II MASTER    20 GOSUB725:IFCO=32THEN30
   1 **                      25 GOSUB450:GOSUB650:I=RI-3:
15 CLS:GOSUB500:GOSUB550:        J=RJ-3:GOSUB725:IFCO<>
   GOSUB600                      32THEN25
```

```
30 GOSUB400:GOTO20

400 REM ** MOVE II-1 **

405 POKENP,32:NP=NP+I+64*J:POKENP,140:RETURN

450 REM ** BLINK II-1 **

455 POKENP,32:POKENP,140:RETURN

500 REM ** HEAD II-1 **

505 CLS:PRINTTAB(25)"ALPHA-II":PRINTTAB(20)"CONVERSION DEMO"
    :FORL=1TO8:PRINT:
NEXT:INPUT"DO 'ENTER' TO START";S$:CLS:RETURN

550 REM ** FIELD II-1 **

555 FORF=15553TO15615:POKEF,131:NEXT

560 FORF=16193TO16255:POKEF,176:NEXT

565 FORF=15553TO16255STEP64:POKEF,191:NEXT

570 FORF=15615TO16255STEP64:POKEF,191:NEXT

575 RETURN

600 REM ** INITIAL II-1 **

605 SF=0:NP=15968:GOSUB650:I=RI-3:J=RJ-3:RETURN

650 REM ** FETCH NEW II-1 **

655 RI=RND(5):RJ=RND(5):IFSF=0ANDRI=3ANDRJ=3THEN655

660 RETURN

725 REM ** CON SENSE II-1 **

730 FORSI=ABS(I)TO1STEP-1:FORSJ=ABS(J)TO1STEP-1

735 CO=PEEK(NP+SI*SGN(I)+64*SJ*SGN(J))

740 IFCO<>32RETURN

745 NEXTSJ,SI:RETURN
```

Alpha-II With Track Sense Option Listing

```
10 REM ** ALPHA-II MASTER 2 **
15 CLS:GOSUB500:GOSUB550:GOSUB600

20 GOSUB725:IFCO=32OR(CO=42ANDTS=0)THEN30

25 GOSUB450:GOSUB650:I=RI-3:J=RJ-3:GOSUB725:IFNOT
   (CO=32OR(CO=42ANDTS=0))THEN2
5

30 GOSUB400:GOTO20
```

134

```
400 REM ** MOVE II-2 **

405 POKENP,42:NP=NP+I+64*J:POKENP,140:RETURN

450 REM ** BLINK II-1 **

455 POKENP,32:POKENP,140:RETURN

500 REM ** HEAD II-2 **

505 CLS:PRINTTAB(25)"ALPHA-II":PRINTTAB(20)"TRACK SENSE OPTION"

510 PRINT:PRINT:PRINT"WANT CREATURE TO BE SENSITIVE
    TO TRACKS":PRINTTAB(10)"0
  - NO":PRINTTAB(10)"1 - YES":INPUTTS

515 CLS:ONTS+1GOTO520,525

520 PRINT@960,"NOT SENSITIVE TO OWN TRACKS":RETURN

525 PRINT@960,"SENSITIVE TO OWN TRACKS":RETURN

550 REM ** FIELD II-1 **

555 FORF=15553TO15615:POKEF,131:NEXT

560 FORF=16193TO16255:POKEF,176:NEXT

565 FORF=15553TO16255STEP64:POKEF,191:NEXT

570 FORF=15615TO16255STEP64:POKEF,191:NEXT

575 RETURN

600 REM ** INITIAL II-1 **

605 SF=0:NP=15968:GOSUB650:I=RI-3:J=RJ-3:RETURN

650 REM ** FETCH NEW II-1 **

655 RI=RND(5):RJ=RND(5):IFSF=0ANDRI=3ANDRJ=3THEN655

660 RETURN

725 REM ** CON SENSE II-1 **

730 FORSI=ABS(I)TO1STEP-1:FORSJ=ABS(J)TO1STEP-1
735 CO=PEEK(NP+SI*SGN(I)+64*SJ*SGN(J))

740 IFCO<>32RETURN

745 NEXTSJ,SI:RETURN
```

Alpha-II With Multi-Sense Printing Program

```
10 REM ** ALPHA-11 MASTER 3 **

15 CLS:GOSUB500:GOSUB550:GOSUB675:GOSUB600
20 GOSUB725:IFCO=32THEN30
```

135

```
25 PRINT@0,CHR$(CO):GOSUB450:GOSUB650:I=RI-3:J=RJ-3:GOSI
   JB725:IFCO<>32THEN25

30 GOSUB400:GOTO20

400 REM ** MOVE II-1 **

405 POKENP,32:NP=NP+I+64*J:POKENP,140:RETURN

450 REM ** BLINK II-1 **

455 POKENP,32:POKENP,140:RETURN

500 REM ** HEAD II-3 **

505 CLS:PRINTTAB(25)"ALPHA-II":PRINTTAB(20)"WITH SENSE
    PRINT"

510 FORL=1TO8:PRINT:NEXT:INPUT"DO 'ENTER' TO START";S$:
    CLS:RETURN

550 REM ** FIELD II-1 **

555 FORF=15553TO15615:POKEF,131:NEXT

560 FORF=16193TO16255:POKEF,176:NEXT

565 FORF=15553TO16255STEP64:POKEF,191:NEXT

570 FORF=15615TO16255STEP64:POKEF,191:NEXT

575 RETURN

600 REM ** INITIAL II-1 **

605 SF=0:NP=15968:GOSUB650:I=RI-3:J=RJ-3:RETURN

650 REM ** FETCH NEW II-1 **

655 RI=RND(5):RJ=RND(5):IFSF=0ANDRI=3ANDRJ=3THEN655

660 RETURN

675 REM ** FIGS II-1 **

677 P=15755

680 FORF=0TO1:FORN=P+64*FTOP+64*F+8:POKEN,35:NEXTN,F

682 FORF=0TO1:FORN=P+32+64*FTOP+40+64*F:POKEN,48:NEXTN,F

684 FORF=0TO1:FORN=P+192+64*FTOP+200+64*F:POKEN,42:NEXTN,F

686 FORF=0TO1:FORN=P+224+64*FTOP+232+64*F:POKEN,37:NEXTN,F

690 RETURN

725 REM ** CON SENSE II-1 **

730 FORSI=ABS(I)TO1STEP-1:FORSJ=ABS(J)TO1STEP-1
```

136

```
735 CO=PEEK(NP+SI*SGN(I)+64*SJ*SGN(J))

740 IFCO<>32RETURN

745 NEXTSJ,SI:RETURN
```

Alpha-II With Sense Responses Listing

```
10 REM ** ALPHA-II MASTER 4 **

15 CLS:GOSUB500:GOSUB550:GOSUB675:GOSUB600

20 GOSUB725:IFCO=32THEN30

22 GOSUB900

25 GOSUB450:GOSUB650:I=RI-3:J=RJ-3:GOSUB725:IFCO<>32THEN
   25

30 GOSUB400:GOTO20

400 REM ** MOVE II-1 **

405 POKENP,32:NP=NP+I+64*J:POKENP,140:RETURN

450 REM ** BLINK II-1 **

455 POKENP,32:POKENP,140:RETURN

500 REM ** HEAD II-4 **

505 CLS:PRINTTAB(25)"ALPHA-II":PRINTTAB(18)" WITH  SENSE
    RESPONSES"

510 FORL=1TO8:PRINT:NEXT:INPUT"DO 'ENTER' TO  START";S$:
    CLS:RETURN

550 REM ** FIELD II-1 **

555 FORF=15553TO15615:POKEF,131:NEXT

560 FORF=16193TO16255:POKEF,176:NEXT

565 FORF=15553TO16255STEP64:POKEF,191:NEXT

570 FORF=15615TO16255STEP64:POKEF,191:NEXT

575 RETURN

600 REM ** INITIAL II-1 **

605 SF=0:NP=15968:GOSUB650:I=RI-3:J=RJ-3:RETURN

650 REM ** FETCH NEW II-1 **

655 RI=RND(5):RJ=RND(5):IFSF=0ANDRI=3ANDRJ=3THEN655

660 RETURN

675 REM ** FIGS II-1 **

677 P=15755

680 FORF=0TO1:FORN=P+64*FTOP+64*F+8:POKEN,35:NEXTN,F
```

137

```
682 FORF=0TO1:FORN=P+32+64*FTOP+40+64*F:POKEN,48:NEXTN,F

684 FORF=0TO1:FORN=P+192+64*FTOP+200+64*F:POKEN,42:NEXTN,F

686 FORF=0TO1:FORN=P+224+64*FTOP+232+64*F:POKEN,37:NEXTN,F
690 RETURN

725 REM ** CON SENSE II-1 **

730 FORSI=ABS(I)TO1STEP-1:FORSJ=ABS(J)TO1STEP-1

735 CO=PEEK(NP+SI*SGN(I)+64*SJ*SGN(J))

740 IFCO<>32RETURN

745 NEXTSJ,SI:RETURN

900 REM ** SENSE RESP II-1 **

905 FORN=0TO40:PRINT@N," ";:NEXT:PRINT@0,CHR$(CO)

910 IFCO=131ORCO=176ORCO=191PRINT@10,"OOF -- A WALL!"

915 IFCO=35ORCO=37PRINT@10,"OUCH -- THAT HURTS!"

920 IFCO=42THEN935

925 IFCO=48THEN940

930 RETURN

935 PRINT@10,"OH, BOY -- FUZZY THINGS":FORTD=1TO250:GOSUB
    450:NEXT:RETURN

940 PRINT@10,"UMM, COOKIES!!":FORTD=1TO1000:NEXT:GOSUB400
    :RETURN
```

Alpha-II With Dual Creatures Program

```
10 REM ** ALPHA-II MASTER 5 **

15 CLS:GOSUB500:GOSUB550:GOSUB600

20 MX=MX*-1:IFMX<0THEN30

25 I=I1:J=J1:NP=M1:GOTO40

30 I=I2:J=J2:NP=M2

40 GOSUB725:IFCO=32THEN55

45 GOSUB900

50 GOSUB650:I=RI-3:J=RJ-3:GOSUB725:IFCO<>32THEN50

55 GOSUB400

60 IFMX<0THEN70

65 I1=I:J1=J:M1=NP:GOTO20
```

138

```
70 I2=I:J2=J:M2=NP:GOTO20

400 REM ** MOVE II-3 **

405 POKENP,32:NP=NP+I+64*J

410 IFMX<0THEN420

415 POKENP,82:RETURN

420 POKENP,66:RETURN

500 REM ** HEAD II-5 **

505 CLS:PRINTTAB(25)"ALPHA-II":PRINTTAB(18)"WITH DUAL
    CREATURES"

510 FORL=1TO8:PRINT:NEXT:INPUT"DO  'ENTER' TO START";
    S$:CLS:RETURN

550 REM ** FIELD II-1 **

555 FORF=15553TO15615:POKEF,131:NEXT

560 FORF=16193TO16255:POKEF,176:NEXT

565 FORF=15553TO16255STEP64:POKEF,191:NEXT
570 FORF=15615TO16255STEP64:POKEF,191:NEXT
575 RETURN
600 REM** INITIAL ii-2 **

605 MX=-1:SF=0:M1=15964:M2=15972

610 GOSUB650:I1=RI-3:J1=RJ-3:GOSUB650:I2=RI-3:J2=RJ-3

615 RETURN

650 REM ** FETCH NEW II-1 **

655 RI=RND(5):RJ=RND(5):IFSF=0ANDRI=3ANDRJ=3THEN655

660 RETURN

725 REM ** CON SENSE II-1 **

730 FORSI=ABS(I)TO1STEP-1:FORSJ=ABS(J)TO1STEP-1

735 CO=PEEK(NP+SI*SGN(I)+64*SJ*SGN(J))

740 IFCO<>32RETURN

745 NEXTSJ,SI:RETURN

900 REM ** SENSE RESP II-2 **

905 IFCO=131ORCO=176ORCO=191RETURN

910 IFMX<0THEN920

915 PRINT@0,"R HIT B":RETURN

920 PRINT@0,"B HIT R":RETURN
```

BETA-I BASIC —
The Next Big Step

There should be no doubt about it, Alpha creatures are fun and instructive. And what's even more important is the fact that they are critical to the evolution of adaptive machine intelligence.

If you have done all the work for the Alpha-Class projects suggested in Chapters 3 through 8, you have learned a lot about machine intelligence, and perhaps you've developed at least a vague notion concerning all the basic groundwork that remains to be done with those simple creatures. However, there has to be much more to adaptive intelligence than random behavior. Somewhere along the line it is necessary to introduce a whole new dimension — memory.

A Beta-Class machine still relies heavily upon elements of Alpha behavior, and throughout the next few chapters, you'll be seeing bits and pieces of Alpha programs and subroutines tucked into the Beta flowcharts and programs. The big difference is that a Beta-Class mechanism couples Alpha-Class behavior with an ability to remember events in the past. And of course, it not only remembers past events, but also has the capacity to act upon those memories at any time in the future. Generally speaking, a Beta-Class mechanism is an Alpha creature that is coupled to a memory scheme.

You'll be starting from ground zero again, but at a new level. Fortunately, many of the subroutines you prepared in earlier chapters can be applied directly to the Beta schemes, thus saving you some time and tedious effort. The point is that you'll be starting out with the most elementary form of Beta mechanism, gradually refining and expanding it through the next few chapters. You are probably in for some surprises and, hopefully, a lot of fun as well.

THE BASIC BETA-I FLOWCHART

All Beta-Class behavior is built upon one rather simple sequence of operations. Study the flowchart in Fig. 9-1 very carefully. If you let things get out of hand here in the early going, there isn't much chance you'll be able to understand and appreciate some of the more interesting, but subtle, features in later versions.

The sequence of program operations for BETA-I BASIC begins by allocating some space for the creature's special memory. That is represented by the first block on the flowchart, SET MEMORY DIM (*dim*ension.) After that, the computer prints a simple heading to identify the program and then the memory is initialized at the block, INITIALIZE MEMORY.

The next two steps, DRAW FIELD and INITIALIZE CRE-ATURE, carry over directly from your ALPHA programs, drawing the same border figure and setting the initial position and direction of the creature. SET NEXT MOVE and CONTACT are

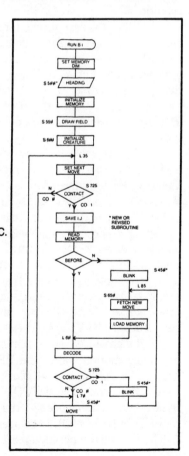

Fig. 9-1. Flowchart for BETA-I BASIC.

likewise identical to ALPHA operations, but some major differences creep into the picture after that. Suppose, however, there is not contact, the CONTACT conditional is *not* satisfied. In that case, the flowchart shows operations going down to a MOVE block which causes the creature to move and subsequently return to SET NEXT MOVE. So once the little creature is underway, it runs through a nice program loop between SET NEXT MOVE, through CONTACT and MOVE. Moving freely across the screen, this BETA creature works just like its simpler ALPHA counterpart.

The differences in behavior become apparent when a contact situation occurs. Satisfying the first CONTACT conditional, the one following SET NEXT MOVE, brings up three new kinds of operations: SAVE I, J, READ MEMORY, and a BEFORE conditional.

SAVE I, J is responsible for saving the values of the motion codes that prevail the moment contact occurs. The BETA scheme, you see, will ultimately assign a particular response to a given contact situation. The response will be stored as data in the system's memory, and the address of that data will be the motion codes that exist the moment the contact situation occurs. This sentence makes a vital point. Mark it so you can locate it easily a bit later in this discussion.

Once the motion codes are stored in memory, actually assigned to other variable names, the next step is to read the memory. What is the address of the memory data to be read? Why, the address specified by the values of the I and J motion codes. A contact situation that occurs while the creature is executing a particular combination of motion codes accesses a unique address in the creature's memory. "I've run into something while moving with a vector I=1 and J=−2. I'll remember those two numbers (SAVE I,J) and then look into my memory to see if I have ever run into something while running I=1 and J=−2 before." That's the sort of "thinking" that goes on through step sequence SAVE I, J, READ MEMORY, and BEFORE.

If the creature has encountered a contact situation under the prevailing motion codes, the BEFORE conditional is satisfied, and the Y output of that block carries operations down to a DECODE block. The DECODE operation is necessary here because the motion codes stored in memory and those generated by FETCH NEW MOVE are specified as integers between 1 and 5. The DECODE operation merely subtracts 3 from those codes, adjusting the interval to the range of integers −2 through +2. Remember the same sort of operation following every FETCH NEW MOVE in the ALPHA programs?

OK, so the creature has encountered a certain kind of contact situation before. The stored response is read and decoded. Then it must be checked for validity by a second CONTACT conditional. If there is no contact, the creature is then free to MOVE, SET NEXT MOVE, etc.

Now, consider this summary of events just described. The creature has been running along freely across the screen, executing a series of stepwise motions determined by the current values of motion codes I and J. But then the creature runs into something on the screen, satisfying the first CONTACT conditional.

The current values of I and J are saved and then used as the address for a Beta-Class memory. The creature examines the contents of the memory at that address. If it sees it has encountered an obstacle while executing that particular motion code before, it pulls the data from memory and decodes it. The data in this case represents the solution to the situation — a pair of encoded motion codes. Those previously stored motion codes are then decoded and tried by the second CONTACT conditional. Assuming they work, they are used as motion codes for the next series of stepwise motions across the screen.

But what happens if the creature encounters an obstacle, addresses its Beta-Class memory according to the current values of I and J, reads the memory, and finds it has *not* encountered a contact situation under those particular conditions? In this case, the BEFORE conditional is *not* satisfied ("I haven't done this before."), and the creature resorts to Alpha-type random behavior.

Following the N output from the BEFORE conditional, you will find a rather familiar loop. The creature's figure is blinked on the screen by BLINK. The system fetches a new, randomly generated set of motion codes and FETCH NEW MOVE, loading the "solution" into the Beta-Class memory. What is the address of that "solution?" The address where the "solution" will reside is determined by the motion codes that exist the moment contact occurs. At the time the creature does a LOAD MEMORY, however, it does not know whether or not the solution is a good one. This last sentence is another one of those very important ones. Mark it for future reference.

At any rate, the system picks a random motion code, stores it in encoded form in the memory, and then decodes it at DECODE before actually trying it. The "solution" is then tried by the second CONTACT conditional. If the "solution" works, CONTACT will not be satisfied and the system goes to MOVE to get the creature moving across the screen again. Naturally, the current motion code is different from the one that existed the moment the originating CONTACT conditional was satisfied.

Here's the situation in capsule form. The creature is moving freely across the screen, executing a series of stepwise motions determined by some values of I and J. It encounters an obstacle and uses the current values of I and J as an address for the Beta-Class memory. The system looks at the contents of the memory and sees it has not encountered an obstacle under those conditions of I and J motions before. It responds by picking a randomly generated set of encoded motion codes, and before testing to see whether or not they will clear the contact condition, the new codes are loaded in memory, thereafter serving as a tentative response to the same sort of contact situation in the future.

The new motion code is then decoded and tested. Since we have assumed in the foregoing discussion that the new codes work, the creature pulls away from the contact situation and begins moving across the screen under the control of the new values of I and J. Remember that these new values of I and J are also stored in Beta-Class memory now. So whenever the creature contacts an obstacle under the same conditions at a later time, the solution is immediately at hand, and there will be no need to resort to a random fetching operation.

There is still one more loop in Fig. 9-1 to consider, the one coming from the Y output of the second CONTACT conditional, going through a BLINK operation, and up to FETCH NEW MOVE. This loop is executed under two different conditions: when a newly selected and randomly generated response doesn't work, and when, for some reason, a previously successful response doesn't work.

Suppose, for instance, that the creature bumps into a contact situation it hasn't dealt with before. The BEFORE conditional is not satisfied and the system resorts to the FETCH NEW MOVE loop. The new move is loaded into memory, decoded, and tried at the second CONTACT conditional. This time assume the new motion code doesn't work — the second CONTACT conditional is satisfied. In that case, the system picks a new random code, loads it into memory, and tries it. The system continues looping around here until it happens across a motion code that does work, and each time it picks a different code, it replaces the old one in memory. The motion code that remains in the memory is the one that works.

Taking a longer view of this BASIC BETA-I program flowchart, you should be able to see that the Beta-Class memory is first wiped clean. The creature begins its life knowing absolutely nothing about how to respond to contact situations. Each contact situation causes the mechanism to check the contents of its memory, at an address determined by the motion code prevailing the moment the contact occurs.

In the early going, none of the memory locations contain any solutions to the contact situation, so the creature must resort to Alpha-like behavior—picking a new motion code at random. When it comes across a motion code that works, it is saved in memory for future reference.

As the creature gains experience with its environment, running into obstacles under a variety of different motion codes, the memory of successful solutions grows and the need to resort to Alpha-type behavior lessens. In principle, at least, the creature eventually reaches a level of experience whereby it has a successful code stored in memory for every possible sort of contact situation. This is an ideal case that is rarely, if ever, achieved in practice.

You can see this memory developing by watching the creature's behavior. When the program is first started, the critter flounders around just like an Alpha creature. But given some experience with its environment, you will note the responses taking on a more rational character. In fact, you will find the creature trying to lock in on a certain pattern of motions. There is more to be said about this kind of observable behavior in the next section of this chapter.

Now, if you have been studying the flowchart in Fig. 9-1 with sufficient care, you might be wondering why previously successful, stored responses are checked for validity. You might ask, "If a given response was successful before, why wouldn't it be successful under the same conditions in the future?"

Indeed, a previously successful and remembered response is checked before it is "approved" and used again. After the READ MEMORY block in Fig. 9-1, the system questions, "Have I done this before?" If the answer is "yes," the implication is that a successful motion code resides in memory. It is decoded and then tested by the second CONTACT conditional. Why test a previously successful response? Why not loop directly from DECODE to MOVE?

The successful responses stored in the Beta-Class memory refer only to the I,J motion codes that prompted those responses in the first place. There is no reference at all to the geometry of the environment. A motion code that happens to get the creature away from a contact situation at one place in the environment's geometry might not work at all at some other place. Fortunately, most contact situations can be solved by more than one motion code. There will always be at least one motion code that is better than the others—one that gets the creature away from the largest percentage of contact situations.

So even though one particular code works under one set of conditions, it might not work (even if the I and J motions are the same) under a different set of environmental conditions. If you were to bypass the doublechecking feature, there's a very good chance the creature would find a contact situation specifying the same memory address, but calling for a different solution than the one residing in memory.

The Beta creature must be allowed to change its mind. By allowing this to happen, it eventually comes up with the best respond. This is adaptive behavior in one of its clearest forms. It sets this sort of machine apart from traditional robot, or parabot, thinking.

Carrying this mind-changing, response-optimizing notion to an extreme, suppose the Beta memory isn't systematically cleared at the beginning of the program. Instead, the INITIALIZE MEMORY step loads the creature's memory with a whole lot of random garbage; it can be anything, really.

In this case, the poor creature is beginning its life with a bunch of screwed-up notions about how to deal with the world around it. Maybe it's like some people you know. A Beta machine, however, owes no special *allegiance* to its preconceived ideas about how to deal with the world.

It will try the garbage responses, but that garbage is gradually replaced with responses that actually work in practice. The garbage initially fed into the memory gets worked out by the very nature of Beta-Class behavior.

Imagine, a machine that cleans up its own act as it goes along. If that doesn't smack of real robotics, I don't know what does.

WHAT TO EXPECT FROM RUNNING BETA-I BASIC

There is no need to show you a diagram of the screen for BETA-I BASIC—it looks exactly like that of ALPHA-I BASIC. It's simply a rectangular border that encloses the little blinking figure of the creature.

When you first run this program, you will see the creature starting from its initial position near the upper left-hand corner of its "playpen." It moves downward and to the right, executing an I=1, J=1 motion code. Under these circumstances, the first border contact will be at the bottom of the border figure. The critter's memory is blank at that moment, so it picks a random response (a new I, J motion code) and goes on from there.

The BETA creature will appear to run just like its ALPHA counterpart for a while. After about 20 to 30 contacts, however, you will be able to see that BETA is locking into a certain pattern of motions. The pattern might be a very simple one and thus rather easy to detect. But on the other hand, the pattern might be exceedingly complex and difficult to follow. In any event, BASIC BETA-I *will* eventually establish a pattern of motion that is peculiar to that one creature. You will probably never see another BETA pick the same pattern. The pattern represents that creature's solution to dealing with the environment. In its own terms, it finds a way to move about most effectively.

In a later chapter you will find some ways to upset the pattern and observe the creature's ability to adapt to the situation by establishing a new one. At this point, however, your job is to see BETA find and lock into its own pattern of motions. It will run that pattern indefinitely. Look for that pattern—that's the important thing here. It's a lot of fun, too. When you get tired of watching the critter showing off its intelligent solution, do a BREAK and RUN the program again.

LOADING BASIC BETA-I

A number of the subroutines specified for this program carry over from BASIC ALPHA-I described in Chapter 3. So load that program from cassette tape before making the modifications and additions specified here.

Here are the subroutines from BASIC ALPHA-I that can be used without any modifications:

S.550 DRAW FIELD (FIELD-1)
S.600 INITIALIZE CREATURE (INITIAL-1)
S.725 CONTACT (CON SENSE-1)
S.650 FETCH NEW MOVE (FETCH NEW-1)

Throughout the ALPHA projects, you wrote BLINK operations into the master program at every point where the creature's figure was to blink and move on the screen. To avoid having to write that long statement line over and over again, create this small subroutine:

```
450 REM ** BLINK–1 **
455 SET(X,Y):SET(X+1,Y):RESET(X,Y):RESET(X+1,Y)
460 RETURN
```

> Variables introduced: None
> Variables required at start: none

This will be called S.450 BLINK from now on.

Then, of course, the HEADING has to be rewritten to reflect the fact that the program is dealing with a Beta-Class system.

```
500 REM ** B HEADING–1 **
505 CLS
510 PRINTTAB(25)"RODNEY BETA-I":PRINT
515 PRINTTAB(25)"BASIC VERSION"
520 FORL=0TO10:PRINT:NEXT
525 INPUT"DO ENTER TO START";S$:RETURN
```

> Variables introduced: L, S$
> Variables required at start: none

Test routine

```
1 GOSUB500
2 CLS:PRINT"OK":INPUTX$:GOTO1
```

That takes care of the subroutines for BASIC BETA-I. There is no need for special discussions since they are quite similar or identical to ground you have covered before. The master program, however, calls for some special consideration.

Although BETA-I BASIC MASTER is similar in many respects to the master program for BASIC ALPHA-1, you will be better off entering it from scratch. Remember to delete line numbers not specified in this program.

```
10 REM ** BETA-I BASIC MASTER **
15 DIMM(5,5,2)
20 GOSUB500
22 CLS:FORN=1TO5:FORM=1TO5:FORD=1TO2:M(M,N,O)=3:NEXT:
   NEXT:NEXT
25 CLS:GOSUB550
30 GOSUB600
```

148

```
35 X=X'+I:Y=Y+J:GOSUB725
40 IFCO=ØTHEN7Ø
45 MI=I+3:MJ=J+3
50 RI=M(MI,MJ,1):RJ=M(MI,MJ,2)
55 IF(RI=3)AND(RJ=3)=-1THEN8Ø
60 I=RI-3:J=RJ-3:GOSUB725
65 IFCO=1THEN95
70 GOSUB45Ø
75 GOTO35
80 GOSUB45Ø
85 GOSUB65Ø
90 M(MI,MJ,1)=RI:M(MI,MJ,2)=RJ:GOTO6Ø
95 GOSUB45Ø:GOTO85
```

The first thing of special note is that the Beta memory is specified as a 3-dimensional array. See the DIM statement in line 15. The array is identified as subscripted variable M, a convention used throughout the BETA work. The first two elements of the array, dimensioned to 5 elements each, represent the address components for I and J respectively. The third element, dimensioned at 2 places, holds the "solution"—the contents of the memory. There are thus 25 Beta memory locations, one for each of the possible combinations of I and J addresses. Each of the 25 memory locations holds two items of information: an I component and a J component for the suggested solution.

Line 15 dimensions this Beta memory space, while line 20 presents the HEADING for the program. After executing the HEADING operation, line 22 is the one that initializes the Beta memory. You'll see that all line 22 does is to run through all 25 memory locations, loading a pair of threes into each one. Recall that an encoded 3 represents a no-motion condition. When decoded, a 3 becomes a 0. So line 22 actually loads the memory with stop codes.

Line 25 clears the HEADING from the screen and calls subroutine 550 to draw the border figure. Line 30 initializes the position and motion of the creature, and line 35 represents the SET NEXT MOVE operation in Fig. 9-1.

The statements are then strictly ALPHA-like down to line 45. In line 45, the current values of I and J are encoded and saved as variables MI and MJ respectively. This is the SAVE I,J operation shown on the flowchart in Fig. 9-1.

In line 50, variables RI and RJ, the same two variables ALPHA uses for encoded information from a random-fetching operation, are picked from the Beta memory. RI, for instance, is picked from address MI,MJ,1. This represents the first of two elements stored as data in address location MI,MJ. And from line 45, you can see that those address components represent the current motion codes. Element RJ is picked in a similar fashion, but it is the second element in that address location.

Line 55 represents the BEFORE conditional. During the initializing process, the memory was loaded with threes. Line 55 is satisfied if RI and RJ are both equal to 3, indicating the system has not responded to this particular situation before. (The system will never select a pair of threes for a response because that represents a stop code, and stop codes are not permitted).

Assuming for the moment that line 55 does not turn up a pair of threes, the motion code read from the Beta memory at line 50 is decoded and tested by line 60 which performs the job of the DECODE and second CONTACT operations in Fig. 9-1.

After that, the operations follow the flowchart in a rather straightforward fashion. The only line calling for more explanation is line 90. Line 90 is the LOAD MEMORY operation. Here the values of RI and RJ are written into their respective places in the Beta memory. Note that the address elements are the same ones established up in line 45. So whatever the solution to the contact situation might be (as represented by RI and RJ), it is written into Beta memory at an address dictated by the motion code prevailing the moment the original contact occurred.

BETA-I BASIC WITH SCORING

The creature created by BETA-I BASIC inevitably establishes a habit pattern of motion, and it might be at least intuitively apparent that the creature's scoring rises well above the chance level once that pattern is established.

The program described here simply provides the scoring figures on the display: NO. CONTACTS, NO. GOOD MOVES, and SCORE. These parameters are identical to those described for ALPHA-I WITH SCORING in Chapter 4. In fact, you can use that program as a starting point for building this one. The screen format is identical to the ALPHA version in Fig. 4-1.

The flowchart for BETA-I BASIC WITH SCORING is shown in Fig. 9-2. It is simply the BETA-I BASIC flowchart (Fig. 9-1) with the scoring operations from Chapter 4 fit into the appropriate places. Since you have already covered this territory before, there is little to be said about it.

The simplest way to get the program loaded is to first enter ALPHA-I WITH SCORING from your cassette tape machine. Then modify S.5ØØ HEADING to read like this:

```
500 REM ** BETA HEADING-2 **
505 CLS
510 PRINTTAB(25)"RODNEY BETA-I":PRINT
515 PRINTTAB(20)"BASIC VERSION WITH SCORING"
520 FORL=0TO10:PRINT:NEXT
525 INPUT"DO ENTER TO START";S$:RETURN
```

Variables introduced: L, S$
Variables required at start: none

Test routine
```
1 GOSUB500
2 CLS:PRINT"OK":INPUTX$:GOTO1
```

Next, add S. 450 BLINK as described earlier in this chapter.
And finally, add the following master program for BETA-I
BASIC WITH SCORING:

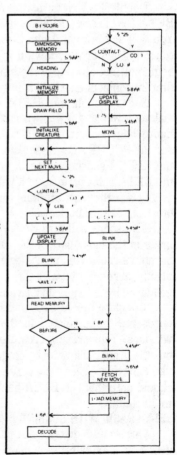

Fig. 9-2. Flowchart for BETA-I BASIC
WITH SCORING.

151

```
10 REM ** BETA-I BASIC SCORE MASTER **
15 DIMM(5,5,2)
17 GOSUB500
20 FORO=1TO2:FORN=1TO5:FORM=1TO5:M(M,N,O)=3:NEXT:NEXT:NEXT
25 CLS:GOSUB550:GOSUB600
30 X=X+I:Y=Y+J:GOSUB725
35 IFCO=0THEN75
40 C=C+1:GOSUB800:GOSUB450
45 MI=I+3:MJ=J+3
50 RI=M(MI,MJ,1):RJ=M(MI,MJ,2)
55 IF(RI=3)AND(RJ=3)=-1THEN80
60 I=RI-3:J=RJ-3:GOSUB725
65 IFCO=1THEN90
70 D=D+1:GOSUB800
75 GOSUB450:GOTO30
80 GOSUB450:GOSUB650
85 M(MI,MJ,1)=RI:M(MI,MJ,2)=RJ:GOTO60
90 C=C+1:GOSUB800:GOTO80
```

Doublecheck your listing against the composite listing for
BETA-I WITH SCORING at the end of this chapter. Then run it.
You will see normal Alpha-Class scoring at first. The score will
run very high or very low through the first few contacts and then
begin settling down to something on the order of .48 through the
first 20 contacts or so. This is typical ALPHA behavior.

Soon, though, the BETA creature finds its habit pattern,
making fewer and fewer wrong moves. The NO. GOOD MOVES
figure begins incrementing right along with NO. CONTACTS,
and the SCORE figure rises slowly, but steadily, upward through
the ALPHA scoring range. If you have become good at spotting
the BETA habit pattern, you will note the rise in scoring as soon
as you see that pattern being executed. The score simply keeps
on rising toward perfection.

The score will never quite reach 1.00 because of the errors
made in the early part of the run, but it rises asymptotically and
far above the levels possible with an ALPHA creature. You can
plainly observe the vast difference in intellectual potential engendered by a simple adaptive memory system.

Compare the ALPHA and BETA programs. The only differences are those related to the memory included in the BETA
versions. Your present creature is essentially Alpha-like at first,
but the Alpha characteristics gradually give way to more Beta-
like behavior—a learning curve that rises toward perfection. You
should note that the BETA behavior begins to dominate after 20
to 30 contacts.

BETA-I BASIC WITH SCORING AND SHOWING MEMORY

One of the distinct advantages of simulating robot intelligence on a computer, as opposed to building a working machine,

is the ease with which the programming can be modified and extended. One of the truly intriguing aspects of BETA-I, as presented on the screen is the ability to observe the action of the memory on a first-hand basis. This little wrinkle can be added to the previous scoring program with little extra effort.

The SHOW MEMORY version works exactly like BETA-I BASIC WITH SCORING. The display is modified, however, to show the status of the Beta memory at any given moment. In this case, the status of the memory is displayed as little spots of light along the bottom of the display screen, each spot having a shape related to the binary value of the information stored there.

There is no need to go into a long dissertation concerning ways to interpret the spots. They can be interpreted, but it is interesting enough just to watch the creature's memory map change in a dynamic fashion as it gains experience with the environment. Once the creature settles into a habit pattern of motion on the screen, you will note that the memory map no longer shows any changes.

So the first thing you should do is save BETA-I BASIC WITH SCORING on cassette tape. Then make the following revisions and additions to the subroutines. Rewrite S.5ØØ HEADING as follows:

```
500 REM ** BETA-I MEM HEAD **
505 CLS
510 PRINTTAB(25)"RODNEY BETA-I":PRINT
515 PRINTTAB(15)"SCORING VERSION WITH SHOW MEMORY"
520 FORL=ØTO1Ø:PRINT:NEXT
525 INPUT"DO ENTER TO START";S$:RETURN
```

Then revise S.55Ø DRAW FIELD. This revision of the standard border figure shortens it a bit from the bottom, thus leaving room for the row of lights representing the Beta memory map.

```
550 REM ** FIELD-2 **
552 F1=15553:F2=15615:F3=16193:F4=16255
555 FORF=F1TOF2:POKEF,131:NEXT
560 FORF=F3TOF4:POKEF,176:NEXT
565 FORF=F1TOF4STEP64:POKEF,191:NEXT
570 FORF=F2TOF4STEP64:POKEF,191:NEXT
575 RETURN
```

The following list of subroutines then carry over without changes from BETA-I BASIC WITH SCORING:

```
S.600 INITIALIZE CREATURE (INITIAL-2)
S.725 CONTACT (CON SENSE-1)
S.800 UPDATE DISPLAY (UD SCORE-1)
S.450 BLINK (BLINK-1)
S.650 FETCH NEW MOVE (FETCH NEW-1)
```

Of course, there has to be a subroutine for displaying the Beta memory map along the bottom of the screen. This is a new one, so it has to be entered from scratch:

```
1000 REM ** SHOW MEM-1 **
1010 W=16327
1015 FORM=1TO5:FORN=1TO5:FORO=1TO2
1020 POKEW,127+M(M,N,O)
1025 W=W+1:NEXT:NEXT:NEXT
1030 RETURN
```

What about a master program for BETA-I BASIC WITH SCORING AND SHOWING MEMORY? That's a simple matter, too. All you have to do is revise line 10 and 85 as follows:

```
10REM ** BETA-I MEM/SCO-1 **
85 M(MI,MJ,1)=RI:M(MI,MJ.2)=RJ:GOSUB1000:GOTO60
```

Check your listing against the composite listing, and you're ready to watch that Beta memory map at work. It's something else!

COMPOSITE PROGRAM LISTINGS

BETA-I BASIC Composite Listing

```
10 REM ** BETA-I BASIC MASTER **

15 DIMM(5,5,2)

20 GOSUB500

22 FORO=1TO2:FORN=1TO5:FORM=1TO5:M(M,N,O)=3:NEXT:NEXT:NEXT

25 CLS:GOSUB550

30 GOSUB600

35 X=X+I:Y=Y+J:GOSUB725

40 IFCO=0THEN70

45 MI=I+3:MJ=J+3

50 RI=M(MI,MJ,1):RJ=M(MI,MJ,2)

55 IF(RI=3)AND(RJ=3)=-1THEN80

60 I=RI-3:J=RJ-3:GOSUB725

65 IFCO=1THEN95

70 GOSUB450

75 GOTO35
```

```
80 GOSUB450

85 GOSUB650

90 M(MI,MJ,1)=RI:M(MI,MJ,2)=RJ:GOTO60

95 GOSUB450:GOTO85

450 REM ** BLINK-1 **

455 SET(X,Y):SET(X+1,Y):RESET(X,Y):RESET(X+1,Y)

460 RETURN

500 REM ** B HEADING-1 **

505 CLS

510 PRINTTAB(25)"RODNEY BETA-i":PRINT

515 PRINTTAB(25)"BASIC VERSION"

520 FORL=0TO10:PRINT:NEXT

525 INPUT"DO ENTER TO START";S$:RETURN

550 REM ** FIELD-1 **

552 F1=15553:F2=15615:F3=16321:F4=16383

555 FORF=F1TOF2:POKEF,131:NEXT

560 FORF=F3TOF4:POKEF,176:NEXT

565 FORF=F1TOF4STEP64:POKEF,191:NEXT

570 FORF=F2TOF4STEP64:POKEF,191:NEXT

575 RETURN

600 REM ** INITIAL-1 **

610 X=10:Y=10:I=1:J=1

620 RETURN

650 REM ** FETCH NEW-1 **

660 RI=RND(5):RJ=RND(5)

665 IFRI=3ANDRJ=3THEN660

670 RETURN

725 REM ** CON SENSE-1 **

730 FORXP=2TO2+ABS(I):FORYP=1TO1+ABS(J)

735 IFPOINT(X+SGN(I)*XP,Y+SGN(J)*YP)=-1THEN750

740 NEXT:NEXT
```

155

```
745 CO=0:RETURN

750 CO=1:RETURN
```

BETA-I BASIC With Scoring

```
10 REM ** BETA-1 BASIC SCORE MASTER **

15 DIMM(5,5,2)

17 GOSUB500

20 FORG=1TO2:FORN=1TO5:FORM=1TO5:M(M,N,G)=3:NEXT:NEXT:NEXT

25 CLS:GOSUB550:GOSUB600

30 X=X+I:Y=Y+J:GOSUB725

35 IFCO=0THEN75

40 C=C+1:GOSUB800:GOSUB450

45 MI=I+3:MJ=J+3

50 RI=M(MI,MJ,1):RJ=M(MI,MJ,2)

55 IF(RI=3)AND(RJ=3)=-1THEN80

60 I=RI-3:J=RJ-3:GOSUB725

65 IFCO=1THEN90

70 D=D+1:GOSUB800

75 GOSUB450:GOTO30

80 GOSUB450:GOSUB650

85 M(MI,MJ,1)=RI:M(MI,MJ,2)=RJ:GOTO60

90 C=C+1:GOSUB800:GOTO80

450 REM ** BLINK-1 **

455 SET(X,Y):SET(X+1,Y):RESET(X,Y):RESET(X+1,Y)

460 RETURN

500 REM ** BETA HEADING-2 **

505 CLS

510 PRINTTAB(25)"RODNEY BETA-i":PRINT

515 PRINTTAB(20)"BASIC VERSION WITH SCORING"

520 FORL=0TO10:PRINT:NEXT

525 INPUT"DO ENTER TO START";S$:RETURN

550 REM ** FIELD-1 **
```

156

```
552 F1=15553:F2=15615:F3=16321:F4=16383

555 FORF=F1TOF2:POKEF,131:NEXT

560 FORF=F3TOF4:POKEF,176:NEXT

565 FORF=F1TOF4STEP64:POKEF,191:NEXT

570 FORF=F2TOF4STEP64:POKEF,191:NEXT

575 RETURN

600 REM ** INITIAL-2 **

610 X=10:Y=10:I=1:J=1

615 C=0:D=0

620 RETURN

650 REM ** FETCH NEW-1 **

660 RI=RND(5):RJ=RND(5)

665 IFRI=3ANDRJ=3THEN660

670 RETURN

725 REM ** CON SENSE-1 **

730 FORXP=2TO2+ABS(I):FORYP=1TO1+ABS(J)

735 IFPOINT(X+SGN(I)*XP,Y+SGN(J)*YP)=-1THEN750

740 NEXT:NEXT

745 CO=0:RETURN

750 CO=1:RETURN

800 REM ** UD SCORE-1 **

805 U$="#.###"

810 E=D/C

850 PRINT@15,"NO. CONTACTS"

852 PRINT@35,"NO. GOOD MOVES"

853 PRINT@55,"SCORE"

855 PRINT@87,C:PRINT@108,D

860 PRINT@119,USINGU$;E

865 RETURN
```

BETA-I BASIC WITH SCORE AND SHOW MEM Listing

```
10 REM ** BETA-I MEM/SCO-1 **

15 DIMM(5,5,2)
```

```
17 GOSUB500

20 FORO=1TO2:FORN=1TO5:FORM=1TO5:M(M,N,O)=3:NEXT:NEXT:NEXT

25 CLS:GOSUB550:GOSUB600

30 X=X+I:Y=Y+J:GOSUB725

35 IFCO=0THEN75

40 C=C+1:GOSUB800:GOSUB450

45 MI=I+3:MJ=J+3

50 RI=M(MI,MJ,1):RJ=M(MI,MJ,2)

55 IF(RI=3)AND(RJ=3)=-1THEN80

60 I=RI-3:J=RJ-3:GOSUB725

65 IFCO=1THEN90

70 D=D+1:GOSUB800

75 GOSUB450:GOTO30

80 GOSUB450:GOSUB650

85 M(MI,MJ,1)=RI:M(MI,MJ,2)=RJ:GOSUB1000:GOTO60

90 C=C+1:GOSUB800:GOTO80

450 REM ** BLINK-1 **

455 SET(X,Y):SET(X+1,Y):RESET(X,Y):RESET(X+1,Y)

460 RETURN

500 REM ** BETA-I MEM HEAD **

505 CLS

510 PRINTTAB(25)"RODNEY BETA-i":PRINT

515 PRINTTAB(15)"SCORING VERSION WITH SHOW MEMORY"

520 FORL=0TO10:PRINT:NEXT

525 INPUT"DO ENTER TO START";S$:RETURN

550 REM ** FIELD-2 **

552 F1=15553:F2=15615:F3=16193:F4=16255

555 FORF=F1TOF2:POKEF,131:NEXT

560 FORF=F3TOF4:POKEF,176:NEXT

565 FORF=F1TOF4STEP64:POKEF,191:NEXT

570 FORF=F2TOF4STEP64:POKEF,191:NEXT
```

```
575 RETURN

600 REM ** INITIAL-2 **

610 X=10:Y=10:I=1:J=1

615 C=0:D=0

620 RETURN

650 REM ** FETCH NEW-1 **

660 RI=RND(5):RJ=RND(5)

665 IFRI=3ANDRJ=3THEN660

670 RETURN

725 REM ** CON SENSE-1 **

730 FORXP=2TO2+ABS(I):FORYP=1TO1+ABS(J)

735 IFPOINT(X+SGN(I)*XP,Y+SGN(J)*YP)=-1THEN750

740 NEXT:NEXT

745 CO=0:RETURN

750 CO=1:RETURN

800 REM ** UD SCORE-1 **

805 U$="#.###"

810 E=D/C

850 PRINT@15,"NO. CONTACTS"

852 PRINT@35,"NO. GOOD MOVES"

853 PRINT@55,"SCORE"

855 PRINT@87,C:PRINT@108,D

860 PRINT@119,USINGU$;E

865 RETURN

1000 REM ** SHOW MEM-1 **

1010 W=16327

1015 FORM=1TO5:FORN=1TO5:FORO=1TO2

1020 POKEW,127+M(M,N,O)

1025 W=W+1:NEXT:NEXT:NEXT

1030 RETURN
```

159

10 Gathering Experimental Data For Beta-I

There is still some serious and significant demonstrations remaining to be described in connection with BETA-I behavior, but the BETA work you've been observing so far ought to be carefully documented first. The habit patterns of motion and steadily increasing scores you've noted from the work in the previous chapter have to be translated into clear and convincing forms, namely some drawings and graphs generated on your own machine.

You will find two basic kinds of data-gathering experiments in this chapter. The first set provides a way to pin down the BETA creatures habit pattern of motion, ultimately giving you a set of clear and unambiguous drawings of those patterns. The second set of experiments compile the results of BETA-I learning activity, providing a learning curve that will serve as a standard for comparison with ALPHA creatures as well as higher-order creatures yet to be described in this book.

SIMPLE TRAIL-DRAWING FEATURE FOR BETA-I

You have probably found that it isn't always easy to detect the habit pattern every BETA creature eventually establishes for itself. The program suggested in this section allows the creature to leave behind a trail wherever it goes, thus greatly simplifying the job of detecting and tracing its habit pattern of motion.

This program is a slight variation of the BETA-I BASIC program already described in great detail in Chapter 9. The trail-drawing feature comes about by making two rather simple modifications. The first modification amounts to omitting the statements in the BLINK subroutine that erases the creature's figure after redrawing it in its next position on the screen. Eliminating the erasing action effectively causes the creature to leave behind an image of itself, a nice trail. However, this first modification will cause some serious problems if it isn't coupled with a second modification. Recall that the BETA creature in Chapter 9 senses potential contact situations by scanning the path ahead of it. Detecting the presence of a lighted segment in its path is tantamount to finding the border figure.

Unfortunately, for our purposes here, the path-scanning feature is also sensitive to the creature's own trail. Upon detecting a section of trail drawn at some earlier time, the creature interprets it as a portion of the border figure and responds by generating a motion code that changes its direction of motion.

Reacting to previously drawn sections of its own trail completely upsets the experiment, creating a dynamic and complex learning situation that masks the elementary features you should be documenting at this time. This might suggest an important project for the future, but it simply messes up the works right now.

The bottom line is that the border sensing scheme has to be changed. The line-drawing program has to be modified so that the BETA creature completely ignores its own trail, but remains sensitive to contacts with the border figure. Now that might sound like a job for BETA-II—a BETA creature that has an extended sensory system that can discriminate between different kinds of objects on the screen. Indeed, BETA-II could handle such a job rather nicely, but you haven't developed such a scheeme yet.

The only alternative is to compromise some of the basic philosophy of good robotics, just long enough to get some pictures of BASIC BETA-I habit patterns. What you will do here is alter the scheme for sensing contact with the border figure, changing it to a position-sensing scheme. Ouch! That hurts the philosophy. No robot worthy of the name should have its behavior dictated by its position relative to a fixed frame of reference. But, maybe you can look at it this way: It is a negative example of good robotics. You did this sort of thing once before when building and using the parabot figure in Chapter 8.

When you begin working with the BETA-II creature you can do the trail-drawing operation without violating any of the fundamental features of good robotics. For now, however, you will be applying a contact-sensing scheme that must not be used as a foundation for creating good theories about evolutionary, adaptive machine intelligence. Just get your pictures of BETA-I trails and let it go at that.

Loading BETA-I WITH TRAIL DRAW

As mentioned earlier, this program is a rather simple variation of the BETA-I BASIC program described in the early part of Chapter 9. So begin the task of loading this trail-drawing program by dumping your BETA-I BASIC into the machine from cassette tape. Then modify S.500 HEADING and add new subroutines S.470 and S.700 as shown here:

```
500 REM**B-I TRAIL HEADING-1**
505 CLS
510 PRINTTAB (25) "RODNEY BETA-I":PRINT
515 PRINTTAB (20)"BASIC VERSION WITH TRAIL"
517 PRINT:PRINT:PRINT"STRIKE 'E' TO ERASE EXISTING TRAIL"
520 FORL=0TO7:PRINT:NEXT
525 INPUT"DO ENTER TO START;S$:RETURN
470 REM**TRAIL DRAW**
475 SET(X,Y):SET(X+1'Y):RETURN
700 REM$$BORD POS SENSE – 1
705 IF(X+1)>124 OR (X+1)<40R(Y+J)>45OR(Y+J)<10THEN 715
710 CO=:RETURN
715 CO=I:RETURN
```

Subroutine S.470 TRAIL DRAW, you will notice, is much the same as S.450 BLINK specified for BETA-I BASIC. This new version, however, does not include the two RESET statements for erasing the creature figure each time it moves to a new position on the screen. This subroutine is thus responsible for both generating the creature figure and its trail.

Subroutine S.700 is the one that causes the philosophical difficulties. It replaces the usual S.725 CONTACT subroutine which scans the path in front of the creature. Instead of looking at the path ahead of the creature, line 705 in subroutine S.700 specifies the exact coordinates of the border figure, thus tying the creature into a fixed frame of reference.

Like S.725 CONTACT, subroutine S.700 returns to its master program carrying a value of CO that indicates whether or not a contact situation is impending. It is thus compatible with other master programs for BETA creatures.

The master program for BETA-I WITH TRAIL DRAW is a simple and direct modification of the master program for BETA-I BASIC. Assuming you have already loaded BETA-I BASIC from cassette tape, you only have to change a few lines. Compare the following listing with that of your BETA-I BASIC, making the necessary changes as you go along.

```
10 REM**BETA-I TRAIL MASTER**
15 DIMM(5,5,2)
20 GOSUB500
22 FORO=1TO2:FORN=1TO5:FORM=1TO5:M(M,N,O)=3NEXT:NEXT:NEXT
25 CLS:GOSUB550
30 GOSUB600
35 X=X+I:Y=Y+J:GOSUB700
40 IFCO=0THEN70
45 MI=I+3:MJ=J+3
50 RI=M(MI,MJ,1):RJ=M(MI,MJ,2)
55 IF(RI=3)AND(RJ=3)=−1THEN80
60 I=RI−3:J=RJ−3:GOSUB70
65 IFCO=1THEN95
70 GOSUB470
75 E$=INKEY$:IFE$<>"E"THEN35
77 CLS:GOSUB550:GOTO35
80 GOSUB470
85 GOSUB650
90 M(MI,MJ,1)=RI:M(MI,MJ,2)=RJ:GOTO60
95 GOSUB470:GOTO85
```

Running the Program and Documenting the Results

Run the program and watch the BETA creature draw its trail on the screen. As long as you can see the creature drawing new paths, you know it has not yet established a complete habit pattern of motion. But once you lose sight of the creature for some period of time, you know it is retracing some sections of the trail pattern, following its own peculiar Beta-type habit.

If you would like to document the complete history of a BETA-I creature (not at all a bad idea from a scientific viewpoint), tape a piece of tracing paper onto the screen, start the program and use a pencil to trace the creature's figure as it moves along. When you lose sight of the creature for an extended period of time, you can be sure it is following its habit pattern and that the trail-drawing job is done. All that, and you have it down on hardcopy, too.

IMPROVED TRAIL-DRAWING PROGRAM FOR BETA-I

The fact that a BETA creature inevitably establishes a habit pattern of motion is an essential feature of the whole scheme of evolutionary machine intelligence and is thus worthy of considerable study and documentation. The trail-drawing program just described goes a long way toward helping you document the habit-forming feature.

The program shows both the creature's habit pattern and the trails it made before full learning took place. That's nice, but the picture can be confusing. The program described in this section allows you to erase the trail drawn prior to the time the creature finally establishes the habit pattern. The result is a clearer picture of the habit pattern. The trade off is the loss of information concerning the learning process.

This revised trail-drawing program also shows the creature's score and, in fact, allows you the option of eliminating the trail-drawing feature altogether. This program, dubbed BETA-I SCORE WITH TRAIL-DRAW OPTION, is an extended version of BETA-I BASIC WITH SCORING described in Chapter 9. The flowcharts are much the same, so there is no need to present a new one in a complete form. To appreciate the differences that do exist, compare the flowchart in Fig. 9-2 with the drawings in Fig. 10-1.

According to Fig. 10-1A, the CONTACT conditionals appearing in Fig. 9-2 are replaced with a set of three conditionals: TRAIL ON, CONTACT, and BORD CONTACT. The critical variable is T, a variable which indicates whether or not the trail is to appear on the screen. If T=1, the trail feature is in effect, but if T=2, the trail is not to be drawn. You can change the value of T, and hence the trail-drawing status, anytime you choose during the execution of the program.

So the TRAIL ON conditional in Fig. 10-1A looks at variable T. If you have specified no trail-drawing operations, TRAIL ON is not satisified and control goes to S.725 CONTACT, the usual path-scanning, contact-sensing subroutine. If you have specified a trail-drawing operation, TRAIL ON is satisfied and the philosophy-blowing kind of border contact sensing (the one already described in the previous section of this chapter) goes into effect.

Whether or not you have opted for the trail drawing feature, the results are the same when operations leave the conditionals in Fig. 10-1A. The system ends up with CO=1 if there is a contact situation and shows a CO=∅ if there is not contact. Everywhere the CONTACT conditional appears on the flowchart in Fig. 9-2, simply replace it with the three conditionals shown in Fig. 10-1A.

Figure 10-1B shows a conditional operation and two ordinary operations which should replace both BLINK and MOVE on the flowchart in Fig. 9-2. If you have opted for the trail-drawing feature, the TRAIL ON conditional in Fig. 10-1B sends operations to S.45∅ TRAIL DRAW, the subroutine that prints the image of the creature, but does not erase the previous image.

If, on the other hand, you have elected to run the program

without showing any of the trail-drawing effects, TRAIL ON in Fig. 10-1A sends operations to the usual BLINK subroutine, S.470. You should recall that this subroutine both draws the creature in its new position and erases the image from the pervious position on the screen.

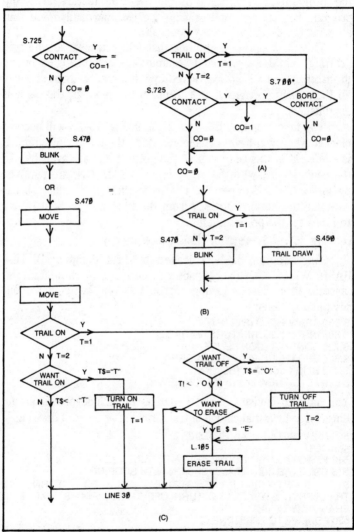

Fig. 10-1. Flowchart modifications for BETA-I SCORE WITH TRAIL-DRAW OPTION (compare with Fig. 9-2) A. Modifications to CONTAC conditionals. B. Modifications for BLINK and MOVE operations. C. Additional operations to be inserted into the flowchart in Fig. 9-2.

The partial flowchart shown in Fig. 10-1C is added to the one shown in Fig. 9-2. This is a set of operations that are inserted between the MOVE block and the line leading back to L.30 in the earlier program. All of these conditionals are intended to give you the option of turning on the trail feature once the program is running, turn off the trail feature after the program has started, and erasing the existing trail figure (but allowing the creature to draw it afresh).

The latter option is the one that lets you clear all the trail drawing from the screen and assuming the creature is executing its habit pattern of motion, it redraws the trail without showing all the messy portions that lead up to the moment the habit is established.

The function of all the operations in Fig. 10-1C will become clearer when you have a chance to study the master program. It is sufficient at this time to note the options that are available. All the conditionals, with the exception of TRAIL ON, are resolved by in-line INKEY$ operations. The implication is that you can exercise the options directly from the keyboard without having to interrupt the program itself.

Loading BETA-I Score With TRAIL-DRAW Option

This program is best loaded by first dumping BETA-I BASIC WITH SCORING (Chapter 9) into your machine from the cassette tape. The following subroutines will be used without any modifications:

S.550 DRAW FIELD (FIELD-1)
S.600 INITIALIZE CREATURE (INITIAL-2)
S.725 CONTACT (CON SENSE-1)
S.800 UPDATE DISPLAY (UD SCORE-1)
S.450 BLINK or MOVE (BLINK-1)
S.650 FETCH NEW MOVE (FETCH NEW-1)

You have to revise S.500 HEADING to include provisions for entering your initial choice of trail drawing or no trail drawing:

```
500 REM**BETA-I TRAIL HEADING-2**
505 CLS
510 PRINTTAB (25) "RODNEY BETA-I":PRINT
515 PRINTTAB (20) "TRAIL VERSION WITH SCORING"
517    PRINT:PRINT:PRINT:"STRIKE    'T'    TO    TURN    ON
TRAIL":PRINT"STRIKE'O' TO TURN OFF TRIAL":PRINT"STRIKE 'E' TO
ERASE ANY TRAIL"
520 FORL= 0 TO5:PRINT:NEXT
525 INPUT"DO ENTER TO START";:RETURN
```

You next will have to enter from scratch S.470 TRAIL DRAW and S. 700 BORD POS SENSE-1 specified in the first section of this chapter.

The master program for BETA-I SCORE WITH TRAIL-DRAW OPTION is an extended version of the master program for BETA-I BASIC WITH SCORING. The latter program should be loaded in program memory now, so all you have to do is search through the following listing, entering new lines and revising a couple of old ones accordingly.

```
10 REM**BETA-I TRAIL SCORE MASTER**
15 DIMM(5,5,2)
17 GOSUB500
20 FOR 0=1TOZ:FORN=1 TO 5:FORM=1 TO 5:M(M,N,O)=3:NEXT:NEX-
    T:NEXT
22 T=1
25 CLS:GOSUB550:GOSUB600
30 X=X+I:Y+J:ONTGOSUB700,725
35 IFCO=0THEN75
40 C=C+1:GOSUB800:ONTGOSUB470,450
45 MI=I+3:MJ=J+3
50 RI=M(MI,MJ,1):RJ=M(MI,MJ,2)
55 IF(RI=3) AND(RJ=3)=-1THEN110
60 I=RI-3:J=RJ-3:ONTGOSUB700,725
65 IFCO=1THEN120
70 D=D+1:GOSUB800
75 ONTGOSUB470,450
80 IFT=1THEN90
85 T$=INKEY$:IFT$="T"THEN87ELSE30
87 T=1:GOTO30
90 T$=INKEY$:IFT$="O"THEN95ELSE100
95 T=2:GOTO105
100 E$=INKEY$:IFE$="E"THEN105ELSE30
105 CLS:GOSUB550:GOTO30
115 M(MI,MJ,1)=RI:M(MI,MJ,2)=RJ:GOTO60
120 C=C+1:GOSUB800 :GOTO10
```

What to Expect From BETA-I SCORE WITH TRAIL-DRAW OPTION

Upon starting this program, you should be greeted with a heading that includes these messages: STRIKE 'T' TO TURN ON TRAIL, STRIKE 'O' TO TURN OFF TRAIL, and STRIKE 'E' TO ERASE ANY TRAIL. These keyboard entries are not to be entered at the beginning of the program, but they can be used anytime after the creature begins moving around the screen.

Line 22 in the master program initializes the value of T to 1. This means the program begins with the trail-drawing feature in effect. If you do not want to see the trail, however, simply strike the 0 key and a moment later the routine will continue without the trail-drawing job.

If, during the course of the program, you want to erase the existing trail, but let the creature continue drawing, strike the E key. A moment later the screen will be completely cleared, you will note the border figure being redrawn, and a moment after that the creature figure and new trail will appear at the point where it left off. The SCORE numbers will be erased too, but they will reappear with the current value as soon as the creature makes contact with the border figure.

If you want to terminate the trail-drawing operation altogether, strike the 0 key. The whole screen will clear, but then the border will be redrawn, the creature will appear in the last position it held, and the score will return as soon as the critter makes contact with the border again.

Incidentally, none of these keyboard changes in this mode of operation can take place while the creature is tangled up in a border-contact situation. Wait until it is running freely across the screen before attempting to change the mode of operation.

Perhaps the most significant application of this program is generating a set of pictures showing the great variety of habit patterns BETA-I creatures can adopt. Start the program, striking the T key as soon as you see the creature moving across the screen. While little BETA is doing its learning operations, tape a sheet of tracing paper over the screen and then wait for the creature to find its habit pattern.

You now have two indicators of a habit pattern. First, you will no longer be able to see the critter, becasue it is simply retracing some old paths. And second, you will note the score rising very steadily.

Once this happens, strike the E key to erase all the trail left during the initial learning process. When the creature's image reappears, begin tracing its path with a soft lead pencil. Use arrows to indicate the direction of motion. If indeed the creature has established its habit pattern of motion, the entire pattern will soon appear quite clearly on the screen. You know the pattern has been completely drawn as soon as you lose sight of the creature for several successive scoring sequences. The experiment and your drawing are then done.

Making a drawing for a new creature is a matter of doing a keyboard BREAK and running the whole operation from the beginning. A collection of six or eight such drawings makes a rather convincing case for the great variety and unpredictable nature of BETA-Class habit formation.

BETA-I WITH SCORE COMPILE AND SHOW MEMORY

The fact that Beta-Class creatures can establish habit patterns of motion is only one of several significant features. You've had a chance to look at some of the habit patterns and watch them develop, now it is time to gather some quantitative data regarding Beta-Class learning curves.

Simply watching the SCORE figures for BETA-I programs can convince you that BETA creatures learn to cope with their environment much more effectively than any ALPHA creature can. Instead of plugging away at an average score of about .46 as ALPHA creatures do, the BETA creatures show a steadily rising learning curve that can approach perfection (1.0) if you let the little character survive long enough.

The procedures and rationale for compiling learning curves has been described before. The most thorough presentation is found in Chapter 6. The program described here operates much the same way, but with a BETA-I creature as the subject. In the early part of the program you specify the number of creatures to be run. For the sake of generating a curve that is statistically reliable, you should specify no less than a hundred creatures. Then once the program begins running the little fellows, you can walk away for a day, letting the computer do the tedious work for you.

More as a matter of casual interest than anything else, this BETA compiling program also shows the Beta-Class memory map along the bottom of the screen.

Figure10-2 shows the printout and learning curve for 100 BETA-I creatures as run according to this program. You will be on firmer scientific grounds if you run your own data, rather than relying on this figure. The learning curve in Fig. 10-2 is clearly different from the one generated by BASIC ALPHA-I in Fig. 6-1. Make sure you understand the significance of the obvious difference between these two curves.

The flowchart for BETA-I WITH SCORE COMPILE AND SHOW MEMORY is drawn in Fig. 10-3. While it appears to be a rather extensive flowchart, a close examination of the details ought to show it is little more than a combination of the flowcharts for BETA-I WITH SCORING (Fig. 9-2) and ALPHA-I COMPILE (FIG.6-3). That's about all this program is, ALPHA-I COMPILE program restructured a bit to deal with BETA creature behavior.

The simplest way to load this program is by first loading ALPHA-I COMPILE (from Chapter 6) into your machine from cassette tape. Then all you have to do is modify or enter a few new subroutines and rewrite the master program. The following subroutines from ALPHA-I COMPILE carry over to this BETA version without modifications:

S.625 INITIALIZE SERIES (INIT COMP-1)
S.650 FETCH NEW MOVE (FETCH NEW-1)
S.725 CONTACT (CON SENSE-1)
S.800 UPDATE DISPLAY (UD SCORE-1)
S.875 COMPILE SERIES (PRINT COMP-1)
S.900 COMPILE RUN (COMPILE-1)

The following subroutines now residing in your program memory
have to be revised:

S.500 HEADING (COMP HEAD-2)
S.550 DRAW FIELD (FIELD-2)
S.600 INITIALIZE CREATURE (INITIAL COMP-3)

The modifications for S.500 are:

```
500 REM**COMP HEAD-2**
505 CLS
510 PRINTTAB (25)"RODNEY BETA-I":PRINT
515 PRINTTAB(20)"BASIC SCORE COMPILE"
517 PRINT:INPUT"HOW MANY TEST CYCLES";T
520 FORL= Ø TO8:PRINT:NEXT
525 INPUT"DO ENTER TO START";S$:RETURN
```

You will find S.550 FIELD-2 and S.600 INITIAL COMP-3
subroutines listed in Chapters 9 and 7 respectively. FIELD-2,
you recall, shortens up the bottom of the border figure allowing
room at the bottom of the screen for printing the Beta memory
map. INITIAL COMP-3 is the subroutine that places the initial
position of the creature at some random spot on the screen,
thereby eliminating the statistical drop in scoring at the beginning of each run. (See the explanation in Chapter 6).

The two remaining subroutines to be added from scratch
are:

S.1000 SHOW MEMORY (SHOW MEM-1)
S.450 BLINK (BLINK1)

Both of these subroutines are listed in Chapter 9.

The master program for BETA-I WITH SCORE COMPILE
AND SHOW MEMORY ought to be entered from scratch — it is
simply too different from its ALPHA counterpart to make a
simple revision possible.

```
10 REM**BETA-I COMPILE MASTER**
15 DIMT(1Ø,2):DIMM(5,5,2)
20 TN=1:GOSUB5ØØ
25 GOSUB625
30 FORO=1TO2:FORN=1TO5:FORM=1TO5:M(M,N,O)=3:NEXTM,N,O
35 CLS:GOSUB55Ø:GOSUB6ØØ
40 X=X+I:Y=Y+J:GOSUB725
45 IFCO=1THEN55
50 GOSUB45Ø:GOTO4Ø
55 C=C+1:GOSUB9ØØ
60 IFC>=THEN13Ø
```

```
65 GOSUB800:GOSUB4500
70 MI=I+3:MJ=J+3
75 RI=M(MI,MJ,1):RJ=M(MI,MJ,2)
80 IF(RI=2)AND(RJ=3)=-1THEN85ELSE90
85 GOSUB450:GOSUB650:M(MI,MJ,1)=RI:M(MI,MJ,2)=RJ:GOSUB1000
90 I=RI-3:J=RJ-3
95 GOSUB725
100 IFCO=1THEN115
105 D=D+1:GOSUB800
110 GOSUB450:GOTO50
115 C=C+1:GOSUB900
120 IFO>=100THEN130
125 GOSUB800:GOT085
130 IFTN<=TTHEN30
135 GOSUB875
140 GOTO20
```

It is a good idea to give the program a couple of brief trial runs before committing it to an extensive series of creature lifetimes. You can start by running the program and responding to the inquiry, HOW MANY CYCLES, by entering the number 1. After the hundred contacts take place (some 15 minutes later), the program should end by printing out the creature's scoring history. Then you should try running the program for 2 creatures, making certain the program shifts to the second one after 100 contacts and properly prints the average of the two histories at the end. If you encounter any problems at all with the program, carefully doublecheck your listing against the one at the end of this chapter.

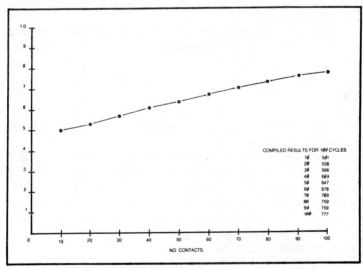

Fig. 10-2. Compiled results of a 100-creature BETA-I experiment.

It takes nearly 24 hours to run and compile the results for 100 creatures. So save this program on cassette tape until you can afford to tie up your computer for that length of time. Copy the results onto a sheet of paper and make the graph for the resulting learning curve. The data can be a valuable source of information for later research. Of course, the most important thing you can do with the results at this time is to compare it with the learning curve you have generated for a hundred ALPHA creatures. There's quite a difference, isn't there?

FOR THE PURISTS: AN ALPHA/BETA COMPILE PROGRAM

Most people would agree that the marked difference between the compiled learning curves for your ALPHA and BETA creatures makes the point. Indeed, BETA creatures show a steadily rising learning curve, while ALPHAs do not. But, there are some critics who might legitimately object to a comparison of the two curves on the grounds that they were run according to two different programs. This is a very fine point, but it has some merit. To get around this sort of objection, you might want to devise a compiling program that shares its mode of operation between ALPHA and BETA creatures. ALPHA creature behavior, you see, is a subset of BETA behavior. All the elements of an ALPHA creature are included in the BETA programs, so it isn't too difficult to modify an existing BETA program to create ALPHA behavior.

The following program gives you the option of running ALPHA or BETA compiling experiments. You can, for instance, set it up to run a hundred creatures in the ALPHA mode, gather the compiled data at the end of the experiment and then set up the same program for running a hundred BETA creatures. The resulting curves thus come from the same program source, and there are no longer any reasonable grounds for questioning the validity of the comparison.

This program is a variation of the BETA-I WITH SCORE COMPILE AND SHOW MEMORY program featured in the previous section of this chapter. So begin the programming process by loading that program into your machine. The following subroutines carry over without any modifications:

S.625 INITIALIZE SERIES (INIT COMP-1)
S.650 FETCH NEW MOVE (FETCH NEW-1)
S.725 CONTACT (CON SENSE-1)
S.800 UPDATE DISPLAY (UD SCORE-1)

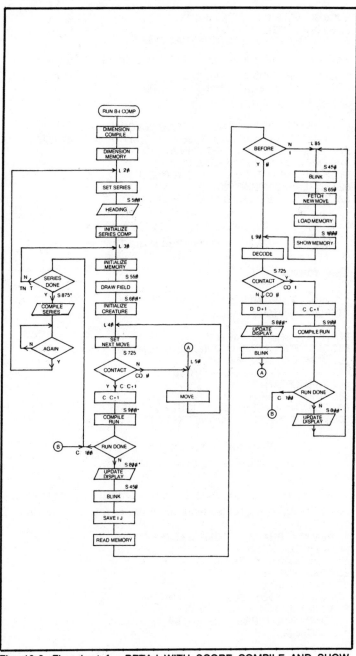

Fig. 10-3. Flowchart for BETA-I WITH SCORE COMPILE AND SHOW MEMORY.

S.900 COMPILE RUN (COMPILE-1)
S.550 DRAW FIELD (FIELD-2)
S.600 INITIALIZE CREATURE (INITIAL COMP-3)

These two subroutines have to be modified a little bit:

S.500 HEADING (COMP HEAD-3)
S.875 COMPILE SERIES (PRINT COMP-1)

```
500 REM ** COMP HEAD-3 **
505 CLS
510 PRINTTAB (20)"RODNEY ALPHA/BETA COMPILE"
515 PRINT:PRINT"ALPHA OR BETA MODE?":PRINTTAB (10)"1 - AL-
PHA":PRI NTTAB(10)"2 - BETA":INPUTMO
516 IFMO=1ORMO=2THEN517ELSE515
517 PRINT INPUT"HOW MANY TEST CYCLES";T
520 FORL=0TO5:PRINT:NEXT
525 INPUT"DO ENTER TO START":S$:RETURN
```

> Variables introduced: MO, L, S$
> Variables required at start: none

Test routine

```
1 GOSUB500:PRINTMO,T
2 INPUTX$:GOTO1

875 REM ** PRINT COMP-2 **
880 CLS:PRINT"COMPILED RESULTS FOR"TN-1"CYCLES":PRINT
881 ONMOGOTO882,883
882 PRINT"ALPHA MODE":GOTO885
883 PRINT"BETA MODE"
885 FORN=1TO10:FC=T(N,1):FD=T(N,2)
890 PRINTN*10"=";:PRINTUSINGU$;FD/FC:NEXT
895 PRINT:PRINT:INPUT"DONE";X$:RETURN
```

> Variables introduced: N, FC, FD, X$
> Variables required to start: U$, TN

The master program need not be completely rewritten. Just make the following revisions or additions to the existing one:

```
10 REM ** ALPHA/BETA COMP MASTER **
36 ONMOGOTO37, 38
37 POKE16325,65:GOTO40
38 POKE16325,66
67 IFMO=1THEN85
85 GOSUB450:GOSUB650:IFMO=1THEN90
87 M(MI, MJ, 1)=RI:M(MI,MJ,2)=RJ
105 D=D+1:GOSUB800:GOSUB1000
110 GOTO50
```

In this master program, MO is a variable that determines whether the program should execute ALPHA or BETA behavior.

Upon running the program, you are greeted with the question, ALPHA OR BETA MODE? 1 - ALPHA, 2 - BETA. If you want to run ALPHA creatures, respond by entering the number 1, and to run a BETA creature, respond by entering the number 2. This INPUT operation sets the value of variable MO.

Whenever you run this program in the ALPHA mode, you will find a letter A printed near the lower left-hand corner of the screen. Running the program in the BETA mode, a letter B is displayed in the lower left-hand corner of the screen, and the BETA memory map is drawn along the bottom of the border figure. When the results are printed out at the end of an experiment, the tabulation is labeled either ALPHA MODE or BETA MODE as appropriate.

The idea is to run the program for at least a hundred creatures, selecting the ALPHA mode during the initial heading phase of the job. Copy down the compiled results some 18 to 20 hours later, and then restart the program, specifying the BETA mode of operation.

The two learning curves gathered from this two-phase project form the basis for a strong argument concerning the evolution of machine intelligence. There is no longer any room for arguing that the BETA program was somehow "loaded" to show the desired results—both curves come from the same program format.

SUGGESTED PROJECT FOR AMBITIOUS EXPERIMENTERS

The two data-compiling programs suggested in this chapter require rather long times for complete execution. It takes something on the order of two days to run both phases of the ALPHA/BETA program.

It is altogether possible to completely rewrite the programs using POKE and PEEK graphics in place of the slower SET and RESET graphics that is specified. The result would be much faster execution of the experiments. You could run a hundred creatures in a much shorter period of time; or, better yet, make it possible to run perhaps a thousand creatures in a tolerable length of time. Remember, the larger the number of creatures you run in a compiling experiment, the more sound your statistical foundation will be.

See the use of PEEK and POKE graphics, as applied to ALPHA-II creatures, in Chapter 8. That should give you some hints concerning the programming procedures.

COMPOSITE LISTINGS
BETA-I WITH TRAIL DRAW Listing

```
10 REM ** BETA-I TRAIL MASTER **
15 DIMM(5,5,2)
20 GOSUB500
22 FORO=1TO2:FORN=1TO5:FORM=1TO5:M(M,N,O)=3:NEXT:NEXT:NEXT
25 CLS:GOSUB550
30 GOSUB600
35 X=X+I:Y=Y+J:GOSUB700
40 IFCO=0THEN70
45 MI=I+3:MJ=J+3
50 RI=M(MI,MJ,1):RJ=M(MI,MJ,2)
55 IF(RI=3)AND(RJ=3)=-1THEN80
60 I=RI-3:J=RJ-3:GOSUB700
65 IFCO=1THEN95
70 GOSUB470
75 E$=INKEY$:IFE$<>"E"THEN35
77 CLS:GOSUB550:GOTO35
80 GOSUB470
85 GOSUB650
90 M(MI,MJ,1)=RI:M(MI,MJ,2)=RJ:GOTO60
95 GOSUB470:GOTO85
450 REM ** BLINK-1 **
455 SET(X,Y):SET(X+1,Y):RESET(X,Y):RESET(X+1,Y)
460 RETURN
470 REM ** TRAIL DRAW **
475 SET(X,Y):SET(X+1,Y):RETURN
500 REM ** B-I TRAIL HEADING-1 **
505 CLS
510 PRINTTAB(25)"RODNEY BETA-i":PRINT
515 PRINTTAB(20)"BASIC VERSION WITH TRAIL"
```

176

```
517 PRINT:PRINT:PRINT"STRIKE 'E' TO ERASE EXISTING TRAIL"

520 FORL=0TO7:PRINT:NEXT

525 INPUT"DO ENTER TO START";S$:RETURN

550 REM ** FIELD-1 **

552 F1=15553:F2=15615:F3=16321:F4=16383

555 FORF=F1TOF2:POKEF,131:NEXT

560 FORF=F3TOF4:POKEF,176:NEXT

565 FORF=F1TOF4STEP64:POKEF,191:NEXT

570 FORF=F2TOF4STEP64:POKEF,191:NEXT

575 RETURN

600 REM ** INITIAL-1 **

610 X=10:Y=10:I=1:J=1

620 RETURN

650 REM ** FETCH NEW-1 **

660 RI=RND(5):RJ=RND(5)

665 IFRI=3ANDRJ=3THEN660

670 RETURN

700 REM ** BORD POS SENSE-1 **

705 IF(X+I)>124OR(X+I)<40R(Y+J)>450R(Y+J)<10THEN715

710 CO=0:RETURN

715 CO=1:RETURN

725 REM ** CON SENSE-1 **

730 FORXP=2TO2+ABS(I):FORYP=1TO1+ABS(J)

735 IFPOINT(X+SGN(I)*XP,Y+SGN(J)*YP)=-1THEN750

740 NEXT:NEXT

745 CO=0:RETURN

750 CO=1:RETURN
```

BETA-I SCORE WITH TRAIL-DRAW OPTION Listing

```
10 REM ** BETA-i TRAIL SCORE MASTER **

15 DIMM(5,5,2)

17 GOSUB500
```

```
20 FORO=1TO2:FORN=1TO5:FORM=1TO5:M(M,N,O)=3:NEXT:NEXT:NEXT
22 T=1
25 CLS:GOSUB550:GOSUB600
30 X=X+I:Y=Y+J:ONTGOSUB700,725
35 IFCO=0THEN75
40 C=C+1:GOSUB800:ONTGOSUB470,450
45 MI=I+3:MJ=J+3
50 RI=M(MI,MJ,1):RJ=M(MI,MJ,2)
55 IF(RI=3)AND(RJ=3)=-1THEN110
60 I=RI-3:J=RJ-3:ONTGOSUB700,725
65 IFCO=1THEN120
70 D=D+1:GOSUB800
75 ONTGOSUB470,450
80 IFT=1THEN90
85 T$=INKEY$:IFT$="T"THEN87ELSE30
87 T=1:GOTO30
90 T$=INKEY$:IFT$="O"THEN95ELSE100
95 T=2:GOTO105
100 E$=INKEY$:IFE$="E"THEN105ELSE30
105 CLS:GOSUB550:GOTO30
110 ONTGOSUB470,450:GOSUB650
115 M(MI,MJ,1)=RI:M(MI,MJ,2)=RJ:GOTO60
120 C=C+1:GOSUB800:GOTO110
450 REM ** BLINK-1 **
455 SET(X,Y):SET(X+1,Y):RESET(X,Y):RESET(X+1,Y)
460 RETURN
470 REM ** TRAIL DRAW **
475 SET(X,Y):SET(X+1,Y):RETURN
500 REM ** BETA-I TRAIL HEADING-2 **
505 CLS
510 PRINTTAB(25)"RODNEY BETA-i":PRINT
515 PRINTTAB(20)"TRAIL VERSION WITH SCORING"
```

```
517 PRINT:PRINT:PRINT"STRIKE 'T' TO TURN ON TRAIL":
    PRINT"STRIKE 'O' TO TURN O
FF TRAIL":PRINT"STRIKE 'E' TO ERASE ANY TRAIL"
520 FORL=0TO5:PRINT:NEXT
525 INPUT"DO ENTER TO START";S$:RETURN
550 REM ** FIELD-1 **
552 F1=15553:F2=15615:F3=16321:F4=16383
555 FORF=F1TOF2:POKEF,131:NEXT
560 FORF=F3TOF4:POKEF,176:NEXT
565 FORF=F1TOF4STEP64:POKEF,191:NEXT
570 FORF=F2TOF4STEP64:POKEF,191:NEXT
575 RETURN
600 REM ** INITIAL-2 **
610 X=10:Y=10:I=1:J=1
615 C=0:D=0
620 RETURN
650 REM ** FETCH NEW-1 **
660 RI=RND(5):RJ=RND(5)
665 IFRI=3ANDRJ=3THEN660
670 RETURN
700 REM ** BORD POS SENSE-1 **
705 IF(X+I)>124OR(X+I)<40R(Y+J)>450R(Y+J)<10THEN715
710 CO=0:RETURN
715 CO=1:RETURN
725 REM ** CON SENSE-1 **
730 FORXP=2TO2+ABS(I):FORYP=1TO1+ABS(J)
735 IFPOINT(X+SGN(I)*XP,Y+SGN(J)*YP)=-1THEN750
740 NEXT:NEXT
745 CO=0:RETURN
750 CO=1:RETURN
800 REM ** UD SCORE-1 **
805 U$="$.###"
```

179

```
810 E=D/C
850 PRINT@15,"NO. CONTACTS"
852 PRINT@35,"NO. GOOD MOVES"
853 PRINT@55,"SCORE"
855 PRINT@87,C:PRINT@108,D
860 PRINT@119,USINGU$;E
865 RETURN
```

BETA-I SCORE COMPILE AND SHOW MEMORY Listing

```
10 REM ** BETA-I COMPILE MASTER **
15 DIMT(10,2):DIMM(5,5,2)
20 TN=1:GOSUB500
25 GOSUB625
30 FORO=1TO2:FORN=1TO5:FORM=1TO5:M(M,N,O)=3:NEXTM,N,O
35 CLS:GOSUB550:GOSUB600
40 X=X+I:Y=Y+J:GOSUB725
45 IFCO=1THEN55
50 GOSUB450:GOTO40
55 C=C+1:GOSUB900
60 IFC>100THEN130
65 GOSUB800:GOSUB450
70 MI=I+3:MJ=J+3
75 RI=M(MI,MJ,1):RJ=M(MI,MJ,2)
80 IF(RI=3)AND(RJ=3)-1THEN85ELSE90
85 GOSUB450:GOSUB650:M(MI,MJ,1)=RI:M(MI,MJ,2)=RJ:GOSUB1000
90 I=RI-3:J=RJ-3
95 GOSUB725
100 IFCO=1THEN115
105 D=D+1:GOSUB800
110 GOSUB450:GOTO50
115 C=C+1:GOSUB900
120 IFC>100THEN130
```

180

```
125 GOSUB800:GOTO85
130 IFTN<=TTHEN30
135 GOSUB875
140 GOTO20
450 REM ** BLINK-1 **
455 SET(X,Y):SET(X+1,Y):RESET(X,Y):RESET(X+1,Y)
460 RETURN
500 REM ** COMP HEAD-2 **
505 CLS
510 PRINTTAB(25)"RODNEY BETA-I":PRINT
515 PRINTTAB(20)"BASIC SCORE COMPILE"
517 PRINT:INPUT"HOW MANY TEST CYCLES";T
520 FORL=0TO8:PRINT:NEXT
525 INPUT"DO ENTER TO START";S$:RETURN
550 REM ** FIELD-1 **
552 F1=15553:F2=15615:F3=16193:F4=16255
555 FORF=F1TOF2:POKEF,131:NEXT
560 FORF=F3TOF4:POKEF,176:NEXT
565 FORF=F1TOF4STEP64:POKEF,191:NEXT
570 FORF=F2TOF4STEP64:POKEF,191:NEXT
575 RETURN
600 REM ** INITIAL COMP-3 **
605 GOSUB650
610 X=RI+25:Y=RJ+25:GOSUB650
612 I=RI-3:J=RJ-3
615 C=0:D=0
620 RETURN
625 REM ** INITIAL COMP-1 **
630 FORN=1TO10:FORM=1TO2:T(N,M)=0:NEXT:NEXT
635 RETURN
650 REM ** FETCH NEW-1 **
660 RI=RND(5):RJ=RND(5)
```

```
665 IFRI=3ANDRJ=3THEN660

670 RETURN

725 REM ** CON SENSE-1 **

730 FORXP=2TO2+ABS(I):FORYP=1TO1+ABS(J)

735 IFPOINT(X+SGN(I)*XP,Y+SGN(J)*YP)=-1THEN750

740 NEXT:NEXT

745 CO=0:RETURN

750 CO=1:RETURN

800 REM ** UD SCORE COMP-1 **

805 U$="#.###"

810 E=D/C

845 PRINT@0,"CREATURE NO."

850 PRINT@15,"NO. CONTACTS"

852 PRINT@35,"NO. GOOD MOVES"

853 PRINT@55,"SCORE"

855 PRINT@87,C:PRINT@108,D

857 PRINT@64,TN" OF"T

860 PRINT@119,USINGU$;E

865 RETURN

875 REM ** PRINT COMP-1 **

880 CLS:PRINT"COMPILED RESULTS FOR"TN-1"CYCLES":PRINT

885 FORN=1TO10:FC=T(N,1):FD=T(N,2)

890 PRINTN*10"=";:PRINTUSINGU$;FD/FC:NEXT

895 PRINT:PRINT:INPUT"DONE";X$:RETURN

900 REM ** COMPILE-1 **

905 IFC/10<>INT(C/10)RETURN

910 TF=INT(C/10):T(TF,1)=T(TF,1)+C:T(TF,2)=T(TF,2)+D

915 IFTF<10RETURN

920 TN=TN+1:RETURN

1000 REM ** SHOW MEM-1 **

1010 W=16327

1015 FORM=1TO5:FORN=1TO5:FORO=1TO2
```

182

```
1020 POKEW,127+M(M,N,O)

1025 W=W+1:NEXT:NEXT:NEXT

1030 RETURN
```

ALPHA/BETA COMPILE Listing

```
10 REM ** ALPHA/BETA COMP MASTER **

15 DIMT(10,2):DIMM(5,5,2)

20 TN=1:GOSUB500

25 GOSUB625

30 FORO=1TO2:FORN=1TO5:FORM=1TO5:M(M,N,O)=3:NEXTM,N,O

35 CLS:GOSUB550:GOSUB600

36 ONMOGOTO37,38

37 POKE16325,65:GOTO40

38 POKE16325,66

40 X=X+I:Y=Y+J:GOSUB725

45 IFCO=1THEN55

50 GOSUB450:GOTO40

55 C=C+1:GOSUB900

60 IFC>100THEN130

65 GOSUB800:GOSUB450

67 IFMO=1THEN85

70 MI=I+3:MJ=J+3

75 RI=M(MI,MJ,1):RJ=M(MI,MJ,2)

80 IF(RI=3)AND(RJ=3)-1THEN85ELSE90

85 GOSUB450:GOSUB650:IFMO=1THEN90

87 M(MI,MJ,1)=RI:M(MI,MJ,2)=RJ

90 I=RI-3:J=RJ-3

95 GOSUB725

100 IFCO=1THEN115

105 D=D+1:GOSUB800:GOSUB1000

110 GOTO50

115 C=C+1:GOSUB900

120 IFC>100THEN130
```

```
125 GOSUB800:GOT085

130 IFTN<=TTHEN30

135 GOSUB875
140 GOTO20

450 REM ** BLINK-1 **

455 SET(X,Y):SET(X+1,Y):RESET(X,Y):RESET(X+1,Y)

460 RETURN

500 REM ** COMP HEAD-3 **

505 CLS

510 PRINTTAB(25)"RODNEY ALPHA/BETA COMPILE"

515 PRINT:PRINT"ALPHA OR  BETA MODE?": PRINTTAB(10)"1
    ALPHA":PRINTTAB(10)"2 - BETA":INPUTMO

516 IFMO=1ORMO=2THEN517ELSE515

517 PRINT:INPUT"HOW MANY TEST CYCLES";T

520 FORL=0TO5:PRINT:NEXT

525 INPUT"DO ENTER TO START";S$:RETURN

550 REM ** FIELD-1 **

552 F1=15553:F2=15615:F3=16193:F4=16255

555 FORF=F1TOF2:POKEF,131:NEXT

560 FORF=F3TOF4:POKEF,176:NEXT

565 FORF=F1TOF4STEP64:POKEF,191:NEXT

570 FORF=F2TOF4STEP64:POKEF,191:NEXT

575 RETURN

600 REM ** INITIAL COMP-3 **

605 GOSUB650

610 X=RI+25:Y=RJ+25:GOSUB650

612 I=RI-3:J=RJ-3

615 C=0:D=0

620 RETURN

625 REM ** INITIAL COMP-1 **

630 FORN=1TO10:FORM=1TO2:T(N,M)=0:NEXT:NEXT

635 RETURN

650 REM ** FETCH NEW-1 **

660 RI=RND(5):RJ=RND(5)

665 IFRI=3ANDRJ=3THEN660

670 RETURN
```

```
 725 REM ** CON SENSE-1 **

 730 FORXP=2TO2+ABS(I):FORYP=1TO1+ABS(J)

 735 IFPOINT(X+SGN(I)*XP,Y+SGN(J)*YP)=-1THEN750
 740 NEXT:NEXT

 745 CO=0:RETURN

 750 CO=1:RETURN
 800 REM ** UD SCORE COMP-1 **

 805 U$="#.###"

 810 E=D/C

 845 PRINT@0,"CREATURE NO."
 850 PRINT@15,"NO. CONTACTS"

 852 PRINT@35,"NO. GOOD MOVES"

 853 PRINT@55,"SCORE"
 855 PRINT@87,C:PRINT@108,D

 857 PRINT@64,TN" OF"T
 860 PRINT@119,USINGU$;E

 865 RETURN
 875 REM ** PRINT COMP-2 **

 880 CLS:PRINT"COMPILED RESULTS FOR"TN-1"CYCLES":PRINT
 881 ONMOGOTO882,883

 882 PRINT"ALPHA MODE":GOTO885
 883 PRINT"BETA MODE"

 885 FORN=1TO10:FC=T(N,1):FD=T(N,2)

 890 PRINTN*10"=";:PRINTUSINGU$;FD/FC:NEXT
 895 PRINT:PRINT:INPUT"DONE";X$:RETURN
 900 REM ** COMPILE-1 **
 905 IFC/10<>INT(C/10)RETURN
 910 TF=INT(C/10):T(TF,1)=T(TF,1)+C:T(TF,2)=T(TF,2)+D
 915 IFTF<10RETURN
 920 TN=TN+1:RETURN
1000 REM ** SHOW MEM-1 **
1010 W=16327
1015 FORM=1TO5:FORN=1TO5:FORO=1TO2
1020 POKEW,127+M(M,N,O)
1025 W=W+1:NEXT:NEXT:NEXT
1030 RETURN
```

11

Disturbing the Beta-I Environment

The discussions, experiments, and programs in Chapter 10 show that a Beta-Class creature seeks a habit pattern of motion and, in the process, exhibits a scoring curve which approaches perfection. Alpha-Class creatures are totally incapable of showing either of these learning features, simply because the lower-class creatures lack a working memory of experiences with the environment.

The work, as presented thus far, does not clearly demonstrate the adaptive behavior of Beta-Class machines. None of the experiments in Chapter 10 suggest upsetting BETA's environment in any significant way, so there was really no opportunity to appreciate its adaptive qualities—the ability to adapt its memory and habit patterns of motion to new and unforseen circumstances in the environment.

The three programs offered in this chapter give BETA a chance to learn its way around the environment, putting successful responses into its memory and eventually establishing a habit pattern of motion. But after that, the environment is upset. The creature is called upon to deal with circumstances not encountered or those dealt with successfully during the initial learning process.

Upsetting the environment after the initial learning phase takes place forces the creature to deal with new circumstances by calling upon successful responses from the past, altering some of the old responses, and, of course, generating and remembering some new ones. This, in a few words, is the adaptive mechanism of Beta-Class creatures at work.

PREVIEW OF BETA-CLASS ADAPTIVE BEHAVIOR

During the initial learning phase of a Beta-Class mechanism, it encounters just so many environmental circumstances. It uses the successful responses to those encounters as guides for dealing with similar situations in the future, and as a result, eventually establishes a pattern of motion that includes many of the successfully executed responses. Not all of the previously successful responses need be included in the habit pattern of motion, but the habit pattern must be built around conditions the creature has dealt with before.

It is highly unlikely that the creature, during the initial learning phase of its existence, will encounter *all* possible environmental conditions. The habit pattern of motion will be developed before the creature has a chance to encounter all possible conditions and use all the responses at its disposal. So once the habit pattern of motion is established, the creature will not be called upon to deal with circumstances it has not encountered before, unless something happens to upset the environment and, hence, the workability of the habit pattern.

There are a number of ways to upset the environment for a BETA creature that is learning its way around on the screen. The simplest and most effective technique, however, is to force upon it a motion code not included in its habit pattern.

All of the creatures described in this book have access to a family of 24 different motion codes. Suppose, during the initial learning process, a BETA creature solves its encounters with the border figure by using 10 of those motion codes and, furthermore, ends up using 8 of them in its habit pattern of motion. Thus, there are 14 motion codes it has never used before and 16 of them that aren't being used in the habit pattern.

Now, if the creature in this example is somehow fed a motion code it has used before, but one that is not included in the habit pattern, the pattern of motion will be temporarily disrupted. In this instance, the creature is presented with a situation it has dealt with before in a successful fashion, and chances are good that the remembered response will work again this time. The pattern will be disrupted for a short time, but it is unlikely the creature will make a bad move. The score will continue to rise since the creature soon picks up its old habit pattern of motion.

But now suppose this same creature is suddenly fed a motion code it has not used before. Upon running into the border figure, the creature must call upon its Alpha-like reflex mechanism to pick a new response. If the new response doesn't work, the score drops a bit, and if the second try doesn't work, the score drops even further. The disturbed creature will eventually find a response that works, remember it, and go about its business. In the meantime, however, the score level has dropped.

How long will it take the BETA creature to find a habit pattern and show a rising score again? That's hard to say, because it depends on how long it takes the creature to get back into a familiar set of circumstances. It might make it to familiar patterns after one move, or it might require a number of encounters (many perhaps entirely new) before setting up a habit pattern again. The new habit pattern of motion might eventually turn out to be identical to the old one, similar to it, or even entirely different.

The point is this: Throwing a BETA creature a motion code it has not executed before disrupts its normal pattern of behavior, forcing it to adapt to a new set of circumstances. The scoring level drops, but as the creature works with the situation, the scoring level begins to rise again and approaches perfection as a habit pattern is worked out.

Recall that Alpha-Class machines are not at all disturbed by changes in their environments. They just keep plugging away at their characteristic, random-oriented scoring level. In a sense, the ALPHA creatures are perfectly adaptable.

Beta-Class machines can be terribly upset, at least temporarily, by new circumstances. They do adapt, but it takes some time. That is a small price to pay, however, for learning scores that surpass anything an Alpha-Class creature can do.

Figure 11-1 compares the scoring of Alpha- and Beta-Class mechanisms before and after a serious disturbance in the environment takes place. Note that the Beta creature is given some time to learn its way around the environment before the disturbance is injected into the scene. BETA shows a rising scoring curve that would run very close to 1.0 if the disturbance weren't injected. ALPHA, however, shows its characteristic, even-level score.

At the moment the disturbance takes place, BETA shows a dramatic drop in its scoring level, while ALPHA remains totally unperturbed. But after the disturbance takes place, BETA's scoring curve begins to rise again. Both creatures have adapted to the disturbance in their own fashion.

There is far more to the adaptive quality of Beta-Class machines that has been described thus far. The discussion to this point, and the curve in Fig. 11-1, apply to a single disturbance following an initial learning phase for the BETA creature. How will BETA respond to a second serious disturbance? A third one? A fourth one?

Each time a Beta-Class machine is presented with a new set of circumstances (and allowed time to deal with the situation) it

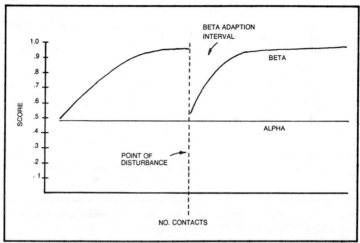

Fig. 11-1. Responses of ALPHA and BETA creatures to a disturbance in the environment.

becomes a bit smarter. Every time BETA is thrust into a new situation, the number of untried responses becomes smaller; or, to put it another way, the number of proven and remembered responses becomes larger.

With experience, gained only by repeated traumas, a Beta-Class machine exhibits improved adaptive qualities. At the very extreme, BETA will end up encountering every conceivable set of environmental circumstances, and there is nothing one can do to the environment to upset the creature. Ideally, it will pick up a quality of perfect adaptibility and a scoring level of 1.00 that cannot be lowered in any way.

Each upset in a BETA's environment makes it a bit smarter and less apt to be thrown into confusion (reaching for random responses) by future changes in the environment. Figure 11-2 shows a hypothetical adaptive curve for a Beta-Class mechanism.

The three programs presented in this chapter give you a chance to observe and test the adaptive qualities of a basic BETA creature. In the first case, you will be able to upset the creature by injecting motion codes of your own selection. The second program slicks up the operation by letting you cycle through all 24 possible motion-code upsets by simply striking the D key.

The third program, the most extensive presented thus far, compiles the adaptive curves for any number of creatures you

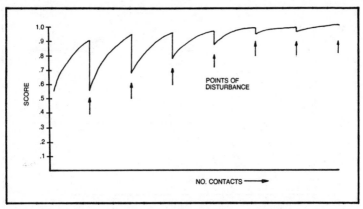

Fig. 11-2. An ideal adaptive learning curve for a BETA creature being subjected to new environmental conditions.

choose. It is fully automatic, and once you set the program into motion, you can forget about using your computer for quite some time. But when the experiment is done, you will have in your hands unquestionable evidence that simple home computers are capable of exhibiting adaptive machine intelligence.

BETA-I ADAPTIVE PROGRAM WITH MANUAL DISTURB

The main purpose of this first program is to let you observe, first hand, the adaptive qualities of a BETA creature. You can enter the "disturbances" at any time and observe the immediate effects. The program also has an ALPHA option, making it easy to compare the results.

Flowchart for Adaptive Program with Manual Disturb

The flowchart for this particular program is illustrated in Fig. 11-3. Notice that it starts out in a rather ordinary way for BETA-I programs: the first step dimensions the memory, a heading is printed on the screen, then the creature's position and direction of motion are initialized, the Beta memory is cleared, and then the field is drawn on the screen.

The only operation of special note to this point is HEADING. It includes a question, ALPHA OR BETA MODE? It's your choice. If you select ALPHA, you will run the program without any sort of memory for the creature, and you cannot expect to see anything new happening whenever you disturb the creature from the keyboard. But, if you opt for BETA, you will be able to see some of the basic characteristics of Beta-Class adaptive behavior whenever you input a disturbance.

All five ALPHA conditionals appearing on the flowchart in Fig. 11-3 are satisfied (Y outputs) if you opt for ALPHA MODE during the HEADING operation. Otherwise, for the BETA MODE, follow the N outputs from each of the ALPHA conditionals.

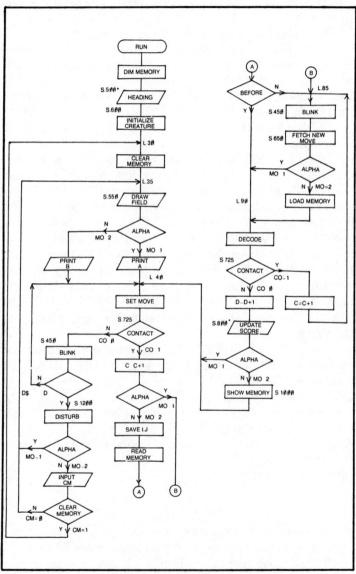

Fig. 11-3. Flowchart for ADAPTIVE PROGRAM WITH MANUAL DISTURB.

191

The first ALPHA conditional, the one immediately following DRAW FIELD simply determines whether an *A* or *B* letter should appear near the lower right-hand corner of the screen. The purpose of the letter is to remind you which mode of operation is in effect. If that ALPHS conditional is satisified, the system prints an *A* down there in the corner, otherwise it prints a *B*.

After printing the appropriate letter designation, the system sets up the first creature move at SET MOVE, and then it scans the path ahead for an obstacle by means of a CONTACT conditional. If that CONTACT conditional is satisfied, the system increments the contact counter at C=C+1 and the program progress in a manner identical to BETA-I BASIC WITH SCORING (FIG. 9-2), with a couple of ALPHA conditionals thrown in as appropriate.

The DISTURB operations are enabled only when the first CONTACT conditional is not satisfied. Return to this particular CONTACT conditional, the one following SET MOVE, and consider the string of operations resulting from a N output. The first operation in that string is a BLINK. There's nothing new about that. It simply blinks the creature figure on and off.

The DISTURB conditional is actually an INKEY$ operation. It is satisfied only if the operator has tapped the *D* key. If the *D* key has *not* been depressed, DISTURB is not satisfied and operations cycle from CONTACT to SET MOVE in the usual fashion—a fashion that allows the creature to move one step on the screen.

But, if the DISTURB conditional is satisfied (because the operator has hit the *D* key), the system calls a DISTURB subroutine. The DISTURB subroutine operation is shown in flowchart form in Fig. 11-4. The first thing the DISTURB subroutine does is print out the current status of the creature's motion code. That's the PRINT STATUS output operation. Actually, the screen is cleared temporarily and you see a message such as:

CURRENT STATUS:
 I=-1
 J=2

This particular example says the creature is currently moving with I=-1 and J=2 (downward and slightly to the left).

The next operation, INPUT NEW I,J, gives you the chance to disturb the creature by entering a new motion code. Maybe

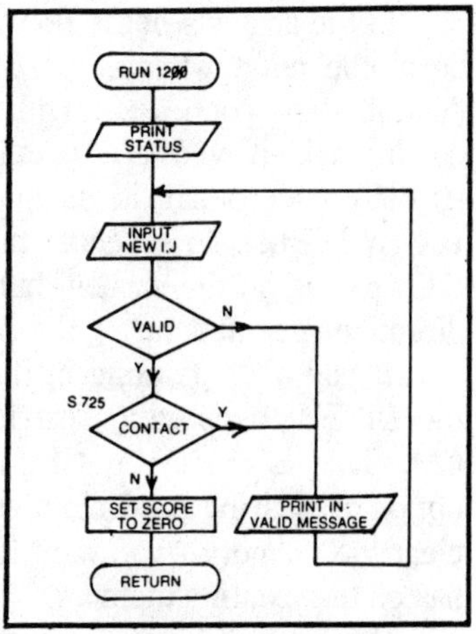

Fig. 11-4. Flowchart for S.12ØØ
DISTURBANCE subroutine.

you'd like it to move straight to the right at full speed. In that case you would enter I=2,J=Ø. You can enter any combination of I and J values you wish, as long as (1) neither value is outside the range of -2 to 2 inclusively, (2) you don't use any fractional expressions such as -1.35, and (3) both values aren't equal to zero. Those three combinations of motion codes are considered invalid, and it is the job of the VALID conditional in Fig. 11-4 to prompt you to enter the right kinds of motion codes. If you should happen to enter an invalid motion code, the system goes to PRINT INVALID MESSAGE and prints INVALID VALUE ENTERED. TRY AGAIN. Operations then move back up to INPUT NEW I,J, thereby giving you a chance to do the job right.

Assuming you enter a valid motion code, the VALID conditional is satisfied, and the next step is to test for contact with the border figure. The scheme as it is set up here cannot accept motion codes of the disturbance variety if those codes run the creature out of the border figure. So, if you enter a motion code that satisfies the CONTACT conditional in Fig. 11-4, you're going to get a polite request for another motion code—presumably one that is both valid and causes no contact with the border figure.

Once you have entered a valid and non-contacting "disturbance," the system sets the score to zero and returns to the main program. By setting the score to zero after doing a disturbance routine, you can see more clearly the effect of the disturbance on the scoring. Without clearing the score, much of the effect would be lost within a past history of successful responses generated during the earlier learning phases.

193

In the analysis of the flowchart in Fig. 11-3, you should now be at the point where the DISTURB subroutine (the one just described in connection with Fig. 11-4) goes to an ALPHA conditional. If you are operating in the ALPHA mode, the ALPHA conditional is satisfied and the program returns to DRAW FIELD. In essence, the original figures are restored to their places on the screen, but now the creature moves off in a direction specified during the DISTURB operation.

If you are operating in the BETA mode, however, leaving the DISTURB operation carries the sequence down to INPUT CM. This is a simple INPUT operation which gives you the option of clearing the Beta memory. Normally you don't want to clear the memory—you want to see what effect your disturbance has on the existing memory.

However, the option of starting a new BETA creature from a blank memory is available at this point. If you choose not to clear the memory, the CLEAR MEMORY conditional is not satisfied and operations go back up to DRAW FIELD to restore the experiment to the point where it was, prior to the DISTURB sequence. Otherwise, operations return all the way to CLEAR MEMORY.

Of course, this program shows the usual array of scoring figures across the top of the screen, and it displays the Beta memory map along the bottom. See the UPDATE SCORE and SHOW MEMORY operations, respectively.

In short, this program allows you to run a basic ALPHA with scoring or a basic BETA with scoring and showing memory. In either case, you can disturb the environment by altering the current motion code and observe the results of the subsequent adaptive intelligence at work.

Loading Adaptive Program With Manual Disturb

The first step for loading this program is to load BETA-I BASIC WITH SCORING AND SHOWING MEMORY from your cassette tape. (See Chapter 9). This will get most of the necessary subroutines into program memory. In fact, the following subroutines carry over without any changes:

S. 550 DRAW FIELD (FIELD-2)
S. 725 CONTACT (CON SENSE-1)
S. 450 BLINK (BLINK-1)
S. 600 INITIALIZE CREATURE (INITIAL-2)
S. 800 UPDATE SCORE (UD SCORE-1)
S. 650 FETCH NEW MOVE (FETCH NEW-1)
S. 1000 SHOW MEMORY (SHOW MEM-1)

Subroutine S.500 HEADING has to be revised a little bit, and, of course, you have to add S.1200 DISTURB (DISTURB-1) from scratch.

```
500 REM ** DIST HEAD-1 **
505 CLS
510 PRINTTAB(20)"RODNEY ALPHA/BETA DISTURB"
515 PRINT:PRINT"ALPHA OR BETA MODE?":PRINTTAB (10)"1 - ALP-
HA":PRINTTAB(10)"2 - BETA":INPUTMO
516 IFMO=1ORMO=2THEN520ELSE515
520 FORL=0TO6:PRINT:NEXT
525 INPUT"DO ENTER TO START";S$:RETURN
```
    Variables required at start: none
    Variables introduced: MO
```
1200 REM ** DISTURB-1 **
1205 CLS:PRINT"CURRENT STATUS:"PRINT
1210 PRINTTAB(10)"=IPRINTTAB(10)"J="J
1215 INPUT"NEW I";I:INPUT "NEW J"
1220 IF(I=0ANDJ=0)OR(I <> INT(I))OR(I<−2ORI ˙ 2)OR(J<>INT(J))OR
     (J < −2ORJ ˙ 2)THEN1230
1222 GOSUB725:IFO=1THEN1230
1225 C=0:D=0:RETURN
1230 PRINT"INVALID VALUE ENTERED. TRY AGAIN":GOTO1215
```

Since the master program in this case has quite a few additional operations, there's nothing to be gained by trying to salvage the old master program. So delete it and enter this one:

```
10 REM ** ALPHA/BETA DISTURB MASTER **
15 DIMM(5,5,2)
20 GOSUB500
25 GOSUB600
30 FORO=1TO2:FORM=1TO5:FORN=1TO5:M(M,N,O)=3:NEXTM,N,O
35 CLS:GOSUB550
36 ONMOGOTO37,38
37 POKE16325,65:GOTO40
38 POKE16325,66
40 X=X+I:Y=Y+J:GOSUB725
45 IFCO=1THEN65
50 GOSUB450:D$=INKEY$:IFD$<>"D"THEN40ELSEGOSUB1200
55 IFMO=1TH
67 IFMO=1THEN85
70 MI=I+3:MJ=J+3
75 RI=M(MI,MJ,1):RJ=M(MI,MJ,2)
80 IF(RI=3)AND(RJ=3)=−1THEN85ELSE90
85 GOSUB450:GOSUB650:IFMO=1THEN90
87 M(MI,MJ,1)=RI:M(MI,MJ,2)=RJ
90 I=RI−3:J=RJ−3
95 GOSUB725
100 IFCO=1THEN120
105 D=D+1:GOSUB800
110 IFMO=2GOSUB1000
115 GOTO40
120 C=C+1:GOTO85
```

195

What to Expect From This Program

Run the program and select the BETA mode of operation. The border figure, creature, and letter *B* should appear on the screen within a short time. The scoring figures and memory map then appear after the creature makes its first contact with the border figure.

Watch the creature's pattern of motion, memory map, and the scoring for signs that it is locking in on its habit pattern. Whenever that happens, the score rises steadily, and the memory map no longer shows any changes whenever the creature makes contact with the border.

With this program, it is possible to doublecheck for this habit-pattern situation. Whenever the creature is moving freely across the screen, strike the D key. This will put the system into the disturb mode of operation. But, you don't want to disturb it right now, so respond to the request for a motion code by entering the one just printed on the screen. This won't change anything as far as the creature's routine is concerned. Respond to the inquiry about clearing the memory by specifying no memory clearing.

After all this, the only difference will be that the score is cleared to zero. And, if the creature has picked up its habit pattern of motion, you will find the score running consistently at 1.00. Entering a motion code that is the same as the current one is always good for resetting the score.

Once the creature has picked up its habit pattern, strike the D key and enter a motion code different from the current one. Do not specify a memory clear. If the creature has encountered your "disturbance" before, the scoring will run at 1.00. You didn't really upset it at all, and it soon picks up its old habit pattern again.

But, if your "disturbance" represents an entirely new motion code, you will see the score starting out at a relatively low point. Yet sometimes it runs at 1.00 for a couple of contacts and then shoots downward. It all depends on how lucky the creature is.

In any event, the creature soon picks up a habit pattern. The new pattern often has the same general characteristics as the old one, tending to move in a clockwise direction, for instance. At other times, the critter picks up the very same pattern. And every once in a while, your disturbance forces the creature to work out an entirely different pattern.

Give the little guy a chance to adjust to your disturbance before hitting it with another. You will see that the creature tends to be disturbed for a shorter duration as you continue disturbing it.

Table 11-1. Phase Number and Corresponding Disturbance Motion Codes.

PHASE NUMBER	DISTURBANCE		PHASE NUMBER	DISTURBANCE	
	I	J		I	J
0	NONE	(INITIAL LEARNING PHASE)			
1	−2	−2	13	0	1
2	−2	−1	14	0	2
3	−2	0			
4	−2	1	15	1	−2
			16	1	−1
5	−2	2	17	1	0
6	−1	−2	18	1	1
7	−1	−1	19	1	2
8	−1	0			
9	−1	1	20	2	−2
			21	2	−1
10	−1	2	22	2	0
11	0	−2	23	2	1
12	0	−1	24	2	2

Keep at it long enough and the creature will not be thrown off the track by any motion code you happen to enter.

You can start out with a fresh creature by either doing a DISTURB operation and responding with a clear memory option, or doing a keyboard BREAK and then running the whole program again from scratch. The only way to switch between ALPHA and BETA creatures is by running the program from the start.

ADAPTIVE BETA-I WITH SEMI-AUTOMATIC DISTURB

This program does essentially the same task as the one just described in the previous section. Rather than entering the disturbance motion codes manually, however, the system automatically selects the codes for you. You still signal the time for a disturbance, but the program cycles the new motion codes through all 24 possibilities for you. When you have cycled through all 24 disturbances, you have the option of clearing the Beta memory and starting all over with a new creature, or restarting the sequence of 24 disturbances with the same creature.

With one exception, the screen format is identical to the one in the previous program. The difference is that this program displays a PHASE number along with the usual scoring figures. The PHASE number indicates which one of the 24 possible disturbances has been executed. Table 11-1 shows the PHASE numbers and their corresponding values of I and J used for upsetting the creature. The program actually involves 25 phases. Phase 0 is the initial learning phase.

Flowchart For Beta-I With Semi-Automatic Disturbance

The flowchart for this program, shown in Fig. 11-5, is practically identical to the manual version in Fig. 11-3. The only real differences are in the DISTURB sequence of operations. The DISTURB conditional is satisfied only when you strike the D key. Otherwise, operations continue as though the disturb feature does not exist.

Upon calling for the DISTURB operations, a counting variable, DT, is incremented at INCREMENT DISTURB PHASE. This counter is initially set to 0 during the initialization part of the program, and it is incremented each time you call for a disturbing operation.

According to the CYCLE DONE conditional, the DISTURB subroutine (S,1200) is called only when DT, the cycle counter, is less than or equal to 24. If the 24th cycle has just been run, the system goes to an INPUT operation asking whether or not you want to proceed with the same memory or a new one. The SAME MEMORY conditional then sends operations to the appropriate point near the beginning of the program. In either case, the phase counter is reset at RESTORE DISTURB CYCLE and the semi-automatic disturbance feature starts all over from PHASE $\emptyset$ again.

The DISTURB subroutine is quite different from its manually operated counterpart in the previous program. This one selects the values of I and J to be forced upon the creature, using the pattern indicated in Table 11-1. Of course, the phase patterns in the DISTURB subroutine are selected to avoid some of the invalid conditions described in connection with the manual version in Fig. 11-4. There is no combination, for instance, where I and J are equal to zero at the same time (motion-stopping code), and none of the figures include fractional parts.

This version of the DISTURB subroutine avoids the problem of injecting motion codes, which cause unwanted contact situations, by resetting the position of the creature on the screen. Each time DISTURB selects a new motion code for testing the creature it also sets the creature figure near the center of the screen—no immediate contact situations are possible.

Some experimenters might question the advisability of disturbing the creature's position as well as its motion code. Studies have shown that tinkering with the creature's position on the screen has little effect on its learning behavior. In fact, it rarely upsets a BETA motion pattern. Such a disturbance might change the relative position of the pattern on the screen, but it does not cause a significant disturbance of the basic learned behavior. The proof and demonstration of this fact is left to you.

So this version of DISTURB automatically picks a new motion code for the creature, sets the creature near the middle of the screen, sets the score to zero, and returns to the main program at a point where the appropriate figures are redrawn on the screen (to DRAW FIELD).

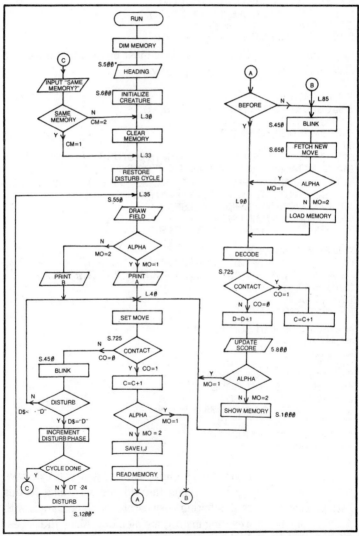

Fig. 11-5. Flowchart for BETA-I WITH SEMI-AUTOMATIC DISTURBANCE.

199

Loading Beta-I With Semi-Automatic Disturbance

This program is loaded using your cassette-tape version of ADAPTIVE PROGRAM WITH MANUAL DISTURB as a starting point. The following subroutines can be used without any modifications whatsoever:

```
S.450 BLINK (BLINK-1)
S.550 DRAW FIELD (FIELD-2)
S.600 INITIALIZE CREATURE (INITIAL-2)
S.650 FETCH NEW MOVE (FETCH NEW-1)
S.725 CONTACT (CON SENSE-1)
S.1000 SHOW MEMORY (SHOW MEM-1)
```

The following three subroutines must be modified or, in the case of S.1200, rewritten completely:

```
S.500 HEADING (DIST HEAD-2)
S.800 UPDATE SCORE (UD DIST SCORE-1)
S.1200 DISTURB (DISTURB-2)
500 REM ** DIST HEAD-1 **
505 CLS
510 PRINTTAB(20)"RODNEY ALPHA/BETA DISTURB"
512 PRINTTAB(20)"(SEMI-AUTOMATIC VERSION)"
515 PRINT:PRINT"ALPHA OR BETA MODE?":PRINTTAB (10)"1 - ALP-
    HA":PRINTTAB(10)"2 - BETA":INPUTMO
516 IFMO=1ORMO=2THEN520ELSE515
520 FORL=0TO6:PRINT:NEXT
525 INPUT"DO ENTER TO START";S$:RETURN

1200 REM ** DISTURB-2 **
1205 -2,-2,-2,-1,-2,0,-2,1,-2,2,-1,-2,-1,-1,-1,0,-1,1,-1,2,0,
    -2,0,-1,0,1,0,2,1,-2,1,-1,1,0,1,1,1,2,2,-1,2,-1,2,0,2,1,2,2
1210 X=25:Y=25
1215 READI,J
1225 C=0:D=0:RETURN

800 REM ** UD DIST SCORE-1 **
805 U$="#.###"
810 E D/C
845 PRINT @ 0, "DISTURB"
850 PRINT @ 15, "NO. CONTACTS"
852 PRINT @ 35, "NO. GOOD MOVES"
853 PRINT @ 55, "SCORE"
855 PRINT @ 87, C:PRINT 100,D
857 PRINT @ 64,DT"OF 24"
860 PRINT @ 119, USINGU$;E
865 RETURN
```

The master programs are quite similar, and you can save yourself a lot of time by revising the old one according to this listing. Read through this listing very carefully, comparing it with the version already residing in program memory. Make the necessary alterations and additions.

```
10 REM ** ALPHA/BETA DISTURB MASTER −2 **
15 DIMM(5,5,2)
20 GOSUB5
25 GOSUB6
30 FORO=1TO2:FORN=1TO5:M(M,N,O)=3:NEXTM,N,O
33 RESTORE:DT=Ø
35 CLS:GOSUB55Ø
36 ONMOGOTO37,38
37 POKE16325,65:GOTO4Ø
38 POKE16325,66
40 X=X+I:Y=Y+J:GOSUB725
45 IFCO=1THEN65
50 GOSUB45Ø:D$=INKEY$:IFD$<>"D"THEN4Ø
55 DT=DT+1:IFDT>24THEN6Ø
57 GOSUB12ØØ:GOTO35
60 INPUT"RECYCLE SAME MEMORY (1=YES,2=NO)";CM:IFCM=1THEN3
    3ELSE3Ø
65 C=C+1
67 IFMO=1THEN85
70 MI≈I+3:MJ=J+3
75 RI=M(MI,MJ,1):RJ=M(MI,MJ,2)
80 IF(RI=3)AND(RJ=3)=-1THEN85ELSE9Ø
85 GOSUB45Ø:GOSUB65Ø:IFMO=1THEN9Ø
87 M(MI,MJ,1)=RI:M(MI,MJ,2)=RJ
9Ø I=RI-3:J=RJ-3
95 GOSUB725
100 IFCO=1THEN12Ø
105 D=D+1:GOSUB8ØØ
110 IFMO=2GOSUB1ØØØ
115 GOTO4Ø
120 C=C+1:GOTO85
```

If you have used the manual version of this program, as
described in the previous section of this chapter, you will find
yourself on familiar ground with this semi-automatic version. The
events proceed in much the same fashion, the only difference being
that the newer version is much easier to use (but you lose control
over which disturbance code is entered when you strike the D key).

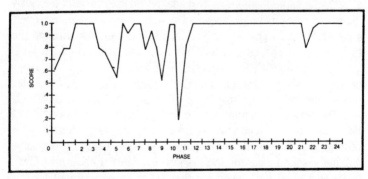

Fig. 11-6. Adaptive curve for one BETA creature subjected to a 24-phase
disturbance cycle.

Getting Some Tangible Results

While it might be fun and interesting (even exciting) to watch the BETA creature adapting to new environmental conditions and learning to adapt more readily with experience, it is difficult to make a good scientific case without some really tangible evidence; this means using the PHASE and SCORE figures to build a graph. Such a graph is shown in Fig. 11-6.

This graph was compiled for a single BETA creature which was run through the initial learning phase and all 24 disturbance test phases. The program was allowed to run for about 25 contacts, and then I struck the D key to cause a disturbance test. After letting the creature adapt for another 25 contacts, I hit the D key again. All along the way, I wrote down the scores at contacts 5 and 25 and then plotted the results.

You will never get the same graph twice, but you ought to consider some of the features of this one. Learning to interpret such graphs is just about as important as making them in the first place.

Notice in Fig. 11-6, for example, that this creature was not at all upset by introducing the motion codes at Phases 1 and 2. It continued learning through Phase 1 and didn't budge from a perfect score of 1.0 through Phase 2. You can see from the figure how the creature adapted and reacted to the various changes. Run a few of these experiments for yourself. It isn't an easy task and it demands your full attention through all 25 phases of the experiment. If you allow 25 contacts between each disturbance test, one complete cycle will require between 45 minutes and an hour.

Since no two BETA creatures deal with their environment in exactly the same fashion, no two adaptation graphs will be exactly the same. After making up several of them, you will note, however, a general trend. The scoring is usually quite erratic through the first 10 or 12 phases and then smooths out at a very high level through the end of the experiment. Every once in a while, though, a critter will show some violent reactions in one or two of the later phases, but you'll find that is an exception to the general trend.

Gathering and compiling the data in this fashion amounts to a lot of time and work on your part. Unless you average together the results of a lot of such experiments, you are going to have trouble making a sound case for the creatures' adaptive qualities. Thus, there is a need for a program that does the experiment for any number of creatures and compiles the results for you.

COMPILING BETA-CLASS ADAPTATION CURVES

The objective of this program is to provide the data necessary for drawing Beta-Class adaptation curves. The job could be done by hand, as described in the previous section of this chapter, but the time and attention required surpasses reasonable bounds. So this program gathers all the data for you, averages the results, and tabulates the scores. All you have to do after that is transfer the tabulated data to a graph.

As in the case of all previous compiling programs, you are asked to enter the number of creatures that are to participate in the experiment. The larger the number of creatures, the more significance the final results will be.

The program initializes a creature, gives it 25 contacts with the border figure to pick up some experience, then it injects the first motion-code disturbance. The creature is allowed to work at the situation for 25 more contacts with the program recording the NO. GOOD MOVES figure at contacts 5 and 25 in that series. The program then injects a different motion-code disturbance, and the creature is allowed to fight its way through another 25 contacts. As before, the number of good moves is recorded at contacts 5 and 25.

This procedure continues through all 24 of the possible motion-code disturbances. NO. GOOD MOVES at contacts 5 and 25 are saved through each one of the 25 disturbance cycles. The score, by the way, is set to zero at the beginning of each new disturbance phase.

So when the creature has been cycled through its initial learning phase and 24 disturbance phases, 48 pieces of data are stored in the compiling memory—NO. GOOD MOVES for contacts 5 and 25 for 24 disturbance phases.

If the system is running more than one creature, the Beta memory is then cleared and a new creature is started. Through this second creature's disturbance cycles its NO. GOOD MOVES is summed with that of the previous creature. After the second creature survives 24 disturbance phases, the Beta memory is cleared and the next creature is created.

This cycle continues until the number of creatures you specified at the beginning of the experiment have all run their course. Then the program processes all the compiled information—the NO. GOOD MOVES summed for all creatures at contacts 5 and 25 through the 24 disturbance phases.

The system divides each of these 48 sums by the number of creatures run in the experiment and prints out the results. The results, in this case, are a set of 48 numbers representing the average creature scores through the 24 disturbance cycles. See an example in Fig. 11-7A. This information is then quite adequate for drawing up the adaptation curve in Fig. 11-7B.

While the curve in Fig. 11-7B appears to be rather erratic, there is a discernable trend in the adaptive ability. As suggested earlier in this chapter, the creatures tend to become more proficient at handling new circumstances as their experience with such circumstances grows. In short, the more you disturb a Beta-Class creature, the better it handles the problems you give it.

Figure 11-7C is a smoothed version of the graph in Fig. 11-7B. The smoothed version rather clearly shows the adaptive trend. Bear in mind, however, that statistical curve-smoothing techniques provide curves that are no more statistically reliable than the original data. And the original data in this case was developed from an experiment involving just 10 creatures.

If you decide to run this compiling experiment for a hundred creatures or more, you can bet that the "unsmoothed" curve would be a lot smoother and more reliable in a statistical sense. There might not even be a need to do a curve-smoothing operation.

So why not run this experiment for a hundred creatures or more? There's just one good reason: Each creature requires between 45 minutes and an hour to go through its paces. This figures out to something on the order of 4 days of around the clock computer time. If you can afford to tie up your computer for a block of time this large, you ought to run the experiment for a hundred creatures.

In any event, the trend in Fig. 11-7 is clear. This business of running the experiment for a hundred creatures is more a philosophical point than a practical one.

Flowchart for Beta Adaptation Compile

The flowchart for this program is shown in Fig. 11-8. It ought to seem rather familiar in a number of respects. You can, for example, find elements of BETA-I BASIC running through it and, of course, there are also some blocks already described in connection with earlier compiling program.

There is something else about this flowchart that might be equally apparent: Many of the standard subroutines are missing. There are no BLINK and SHOW MEMORY operations, for example. When you get a chance to look over the listings, you will find that the usual subroutines, in many cases, have been greatly shortened.

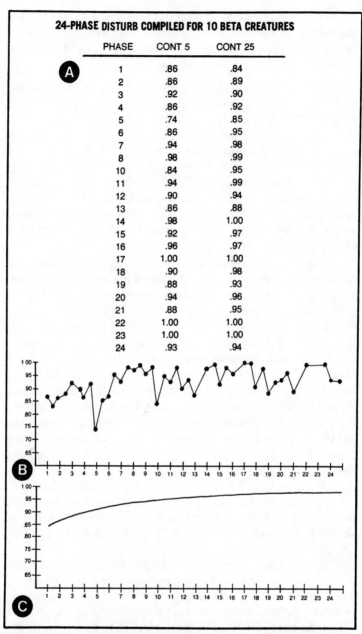

24-PHASE DISTURB COMPILED FOR 10 BETA CREATURES

PHASE	CONT 5	CONT 25
1	.86	.84
2	.86	.89
3	.92	.90
4	.86	.92
5	.74	.85
6	.86	.95
7	.94	.98
8	.98	.99
10	.84	.95
11	.94	.99
12	.90	.94
13	.86	.88
14	.98	1.00
15	.92	.97
16	.96	.97
17	1.00	1.00
18	.90	.98
19	.88	.93
20	.94	.96
21	.88	.95
22	1.00	1.00
23	1.00	1.00
24	.93	.94

Fig. 11-7. Results of running 10 BETA creatures through the BETA ADAPTION COMPILE program. A. Tabulated results from the computer. B. Adaption curve resulting from plotting the raw results. C. Adaptation curve resulting from curve-smoothing the same results.

The reason for skipping some of the standard subroutines and shortening up some of the others is rather simple. This is an extensive program, and it is necessary to cut out a lot of operations in order to get it to fit into a 4K memory.

The SHOW MEMORY was one of the first subroutines that had to be deleted. It never really played a vital role in previous programs, and it would not contribute anything significant here, either. A somewhat more costly trade-off concerns the elimination of the BLINK subroutine. The purpose of this subroutine has been to show the creature figure on the screen. You don't really have to see the creature; it will perform just as well if it is not shown. And since the BLINK subroutine takes up some memory space that is badly needed, it must go. Similar deletions applied within the more necessary subroutines will become apparent when you have to load them.

Since you have worked with compiling programs and BETA-I routines before, there is no need to describe this flowchart in great detail. Some of the variables ought to be defined, though. Variable C, as usual, is the contact counting variable, and variables I and J are on-line motion-code variables. The number of good moves is stored as variable D. There's nothing new about any of these.

DT is the disturb phase cycle counter. Whenever DT=∅, for example, the creature is executing its initial learning phase. And when DT=1, it means the system is running the first disturb phase. Note that the value of DT is incremented in the block labeled DT=DT+1. The disturb phase is thus incremented at that point, and the conditional DT *greater than* 24 looks for the final disturb phase. If the conditional operation is *not* satisfied, it means the disturbing operations aren't complete, and you will see that operations return to DRAW FIELD to begin a new disturb phase.

TN is the creature counter. Each time the system completes its series of 24 disturb cycles (conditional DT *greater than* 24 is satisfied), TN is incremented to assign a number to the next creature in the experiment.

Variable CN is the number of creatures you specify to be run in the experiment. That is actually done way back in the HEADING operation. Whenever TN exceeds the value of CN, it is time to terminate the runs and tabulate the data. Otherwise, the value of TN is assigned to the next creature, TN is initialized at IN-ITIALIZE CREATURE, the previous creature's memory is cleared out at CLEAR MEMORY, and a new baby is launched into the world.

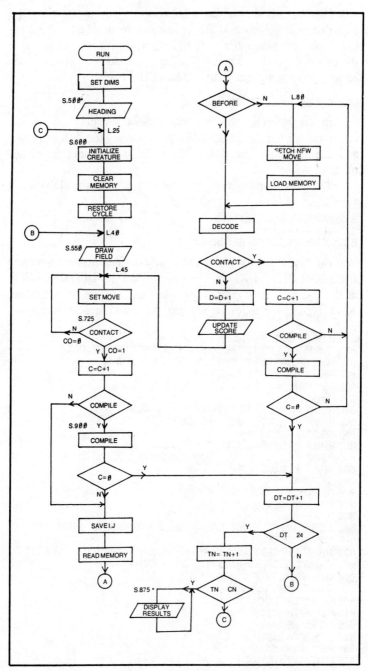

Fig. 11-8. Flowchart for BETA ADAPTATION COMPILE.

Figure 11-9 shows the screen format during the time the experiment is being run. CREAT. shows the number of the creature being run at the moment (TN) and the total number to be run in the experiment (CN). The PHASE figure indicates which one of the 25 phases is being run at the moment (DT).

Loading Beta Adaptation Compile

In this particular case, you will come out ahead by loading this entire program from scratch. There are simply too many changes and additions to make it feasible to build it around an earlier program.

These three subroutines are the only ones identical to earlier versions:

```
S.600 INITIALIZE CREATURE
S.550 DRAW FIELD
S.725 CONTACT (CON SENSE-1)
```

You can find the listings for these subroutines in the composite listings for BETA-I WITH SEMI-AUTOMATIC DISTURBANCE at the end of this chapter. Load them as they are shown there. Next, type in the following 5 subroutines and master program:

```
500 REM ** DIST HEAD-3 **
510 CLS:PRINTTAB(15)"RODNEY BETA DISTURB COMPILE"
517 INPUT"HOW MANY CREATURES";CN:RETURN
800 REM ** UD DIST SCORE-1 **
805 U$="#.###"
810 E=D/C
815 PRINT @ 0 ,"CREAT.":PRINT @ 10,"PHASE"
850 PRINT @ 18,"NO. GOOD MOVES"
853 PRINT @ 55,"SCORE"
855 PRINT @ 87,C:PRINT 108,D
857 PRINT @ 64,TN"OF"CN:PRINT @ 74,DT
860 PRINT @ 119,USINGU$;E
865 RETURN
875 REM ** DIST COMP DATA-1 **
877 TN=TN-1:U$="#.##"
880 DT=1
886 CLS:PRINTTAB(25)"LONG-TERM AVERAGES":PRINT"PHASE",
    "CONT 5", CONT 25 "
890 PRINT:PRINTDT;:PRINTTAB916)USINGU$; (T(DT,1)+ TN/ (TN * 5);:
    PRINTTAB(32)USINGU$;(T(DT,2)+TN)/(TN*25)
892 INPUTS$:DT=DT+1:IFDT >
    24THEN880ELSE886
900 REM ** DIST COMP-1 **
915 IFDT=0THEN945
920 IFC=5THEN935
925 IFC=25THEN940
930 IFC < 25RETURN
935 T(DT,1)=T(DT,1)+D:
    RETURN
```

```
 940 T(DT,2)=T(DT, 2)+D
 945 GOSUB1200:RETURN
1200 REM ** DISTURB-2 **
1205 DJ=DJ+1:IFDJ=0 ANDDI= 0THEN1205
1210 IFDJ 2THEN1215ELSE1220
1215 DJ=-2:DI=DI+1
1220 I=DI:J=DJ:X=25:Y=25
1225 C=0:D= 0:RETURN
```

For the sake of future reference, these subroutines will be called:

```
S.500 HEADING (DIST HEAD-3)
S.800 UPDATE SCORE (UD DIST SCORE-1)
S.875 DISPLAY RESULTS (DIST COMP DATA-1)
S.900 COMPILE (DIST   COMP-1)
```

The master program is as follows:

```
10 REM ** BETA DISTURB COMPILE MASTER **
15 DIMM(5,5,2):DIMT(24,2)
20 TN=1:GOSUB500
25 GOSUB600
30 FORO=1TO2:FORN=1TO5:FORM= TO5:M(M,N,O)=3:NEX TM,N,O
35 DT=0:DI=-2:DJ=-3
40 CLS:GOSUB550
```

Fig. 11-9. Screen format while running BETA ADAPTION COMPILE.

```
45 X=X+I:Y=Y+J:GOSUB725
50 IFCO=1THEN6Ø
55 GOTO45
60 C=C+1:IF(DT=ØAND C > =25) OR (DT > ØAND((C=50RC =25))THEN
   65ELSE7Ø
65 GOSUB9ØØ:IFC=0THEN11Ø
70 MI=I+3:MJ=J+3:RI=M(MI,MJ,1):RJ=M(MI,MJ,2)
75 IFNOT(RI=3ANDRJ=3)THEN85
80 GOSUB65Ø:M(MI,MJ,1)=RI:M(MI,MJ,2)=RJ
85 I=RI-3:J=RJ-3:GOSUB725
90 IFCO=1THEN1ØØ
95 D=D+1:GOSUB'8ØØ:GOTO45
100 C=C+1:IF(DT=ØANDC>=25)OR(DT >ØAND ((C=5ORC=25))THEN
    1Ø5ELSE8Ø
105 GOSUB9ØØ:IFC <> ØTHEN8Ø
110 DT=DT+1:IFDT<=24THEN4Ø
115 TN=TN+1:IFTN<=CNTHEN25
120 GOSUB875
```

Preliminary Tests for Beta Adaptation Compile

This program ought to be thoroughly checked before committing it to a long-term experimental run. The slightest error can
mean many, many hours of wasted computer time.

Start out by running the program and entering a 1-creature
experiment. If all goes well, you should see the phase numbers
changing each time the NO. CONTACTS figure reaches 25. About
an hour later (after the 24th cycle is done), you ought to see this
message on the screen:

	LONG-TERM AVERAGES	
PHASE	CONT 5	CONT 25
1	Ø.85	Ø.95
?—		

There is not enough memory space available in a 4K personal
computer system to bring out the entire results at one time; so
you see the results, one phase at a time. This example shows the
results for PHASE 1. In this particular case, the SCORE at contact
5 (CONT5) is 0.85, and the SCORE at the 25th contact was 0.95.

To see the results for PHASE 2, and all subsequent phases,
simply strike the ENTER key. Incidentally, the display returns to
Phase 1 after showing Phase 24. This little 'round-and-'round
feature lets you doublecheck all the information without running
the risk of loosing it by accident.

If it all works out, run the experiment again, entering 2
creatures this time. The idea is to make sure the system averages
the results properly. Then take a break for a couple of hours while
the program does its job.

You know the testing phase is done when the averages for Phase 1 appear on the screen. Check through all the averages, making sure they are "reasonable." None of the averages should exceed 1.00, for instance, and, of course, there should be no negative numbers or, for that matter, any scores much lower than 0.30 or so.

COMPOSITE LISTINGS

Adaptive Program With Manual Disturb Listing

```
10 REM ** ALPHA/BETA DISTURB MASTER **
15 DIMM(5,5,2)
20 GOSUB500
25 GOSUB600
30 FORO=1TO2:FORN=1TO5:FORM=1TO5:M(M,N,O)=3:NEXTM,N,O
35 CLS:GOSUB550
36 ONMOGOTO37,38
37 POKE16325,65:GOTO40
38 POKE16325,66
40 X=X+I:Y=Y+J:GOSUB725
45 IFCO=1THEN65
50 GOSUB450:D$=INKEY$:IFD$<>"D"THEN40ELSEGOSUB1200
55 IFMO=1THEN35
60 INPUT"WANT NEW MEMORY (1=YES,0=NO)";CM:IFCM=1THEN30ELSE35
65 C=C+1
67 IFMO=1THEN85
70 MI=I+3:MJ=J+3
75 RI=M(MI,MJ,1):RJ=M(MI,MJ,2)
80 IF(RI=3)AND(RJ=3)-1THEN85ELSE90
85 GOSUB450:GOSUB650:IFMO=1THEN90
87 M(MI,MJ,1)=RI:M(MI,MJ,2)=RJ
90 I=RI-3:J=RJ-3
95 GOSUB725
100 IFCO=1THEN120
```

```
105 D=D+1:GOSUB800

110 IFMO=2GOSUB1000

115 GOTO40

120 C=C+1:GOTO85

450 REM ** BLINK-1 **

455 SET(X,Y):SET(X+1,Y):RESET(X,Y):RESET(X+1,Y)
460 RETURN

500 REM ** DIST HEAD-1 **

505 CLS

510 PRINTTAB(20)"RODNEY ALPHA/BETA DISTURB"

515 PRINT:PRINT"ALPHA OR BETA MODE?":PRINTTAB(10)"1
    - ALPHA":PRINTTAB(10)"2 - BETA":INPUTMO

516 IFMO=1ORMO=2THEN520ELSE515

520 FORL=0TO6:PRINT:NEXT

525 INPUT"DO ENTER TO START";S$:RETURN

550 REM ** FIELD-1 **

552 F1=15553:F2=15615:F3=16193:F4=16255

555 FORF=F1TOF2:POKEF,131:NEXT

560 FORF=F3TOF4:POKEF,176:NEXT

565 FORF=F1TOF4STEP64:POKEF,191:NEXT

570 FORF=F2TOF4STEP64:POKEF,191:NEXT

575 RETURN

600 REM ** INITIAL COMP-3 **

605 GOSUB650

610 X=RI+25:Y=RJ+25:GOSUB650

612 I=RI-3:J=RJ-3

615 C=0:D=0

620 RETURN

650 REM ** FETCH NEW-1 **

660 RI=RND(5):RJ=RND(5)

665 IFRI=3ANDRJ=3THEN660

670 RETURN
```

212

```
725 REM ** CON SENSE-1 **

730 FORXP=2TO2+ABS(I):FORYP=1TO1+ABS(J)

735 IFPOINT(X+SGN(I)*XP,Y+SGN(J)*YP)=-1THEN750

740 NEXT:NEXT
745 CO=0:RETURN

750 CO=1:RETURN

800 REM ** UD SCORE-1 **

805 U$="#,###"

810 E=D/C

850 PRINT@15,"NO. CONTACTS"

852 PRINT@35,"NO. GOOD MOVES"

853 PRINT@55,"SCORE"

855 PRINT@87,C:PRINT@108,D

860 PRINT@119,USINGU$;E

865 RETURN

1000 REM ** SHOW MEM-1 **

1010 W=16327

1015 FORM=1TO5:FORN=1TO5:FORO=1TO2

1020 POKEW,127+M(M,N,O)

1025 W=W+1:NEXT:NEXT:NEXT

1030 RETURN

1200 REM ** DISTURB-1 **

1205 CLS:PRINT"CURRENT STATUS:":PRINT

1210 PRINTTAB(10)"I="I:PRINTTAB(10)"J="J:PRINT:PRINT

1215 INPUT"NEW I";I:INPUT"NEW J";J

1220 IF(I=0ANDJ=0)OR(I<>INT(I))OR(I<-20RI>2)OR(J<>INT(J))
     OR(J<-20RJ>2)THEN123 0
1222 GOSUB725:IFCO=1THEN1230

1225 C=0:D=0:RETURN

1230 PRINT"INVALID VALUE ENTERED. TRY AGAIN":GOTO1215
```

Beta-I With Semi-Auto Disturbance Listing

```
10 REM ** ALPHA/BETA DISTURB MASTER **

15 DIMM(5,5,2)
```

```
20 GOSUB500

22 DC=0

25 GOSUB600

30 FORO=1TO2:FORN=1TO5:FORM=1TO5:M(M,N,O)=3:NEXTM,N,O

33 RESTORE

34 FORDT=1TO24

35 CLS:GOSUB550

36 ONMOGOTO37,38

37 POKE16325,65:GOTO40

38 POKE16325,66

40 X=X+T:Y=Y+J:GOSUB725

45 IFCO=1THEN65

50 GOSUB450:D$=INKEY$:IFD$<>"D"THEN40ELSEGOSUB1200

55 NEXTDT

60 INPUT"RECYCLE SAME MEMORY ( 1=YES,2=NO)";CM:IFCM=1
   THEN33ELSE30

65 C=C+1

67 IFMO=1THEN85

70 MI=I+3:MJ=J+3

75 RI=M(MI,MJ,1):RJ=M(MI,MJ,2)

80 IF(RI=3)AND(RJ=3)-1THEN85ELSE90

85 GOSUB450:GOSUB650:IFMO=1THEN90

87 M(MI,MJ,1)=RI:M(MI,MJ,2)=RJ

90 I=RI-3:J=RJ-3

95 GOSUB725

100 IFCO=1THEN120

105 D=D+1:GOSUB800

110 IFMO=2GOSUB1000

115 GOTO40

120 C=C+1:GOTO85

450 REM ** BLINK-1 **

455 SET(X,Y):SET(X+1,Y):RESET(X,Y):RESET(X+1,Y)
```

```
460 RETURN

500 REM ** DIST HEAD-1 **

505 CLS

510 PRINTTAB(20)"RODNEY ALPHA/BETA DISTURB"

515 PRINT:PRINT"ALPHA OR BETA MODE?":PRINTTAB(10)"
    1 - ALPHA":PRINTTAB(10)"2 -  BETA":INPUTMO

516 IFMO=1ORMO=2THEN520ELSE515

520 FORL=0TO6:PRINT:NEXT

525 INPUT"DO ENTER TO START";S$:RETURN

550 REM ** FIELD-1 **

552 F1=15553:F2=15615:F3=16193:F4=16255

555 FORF=F1TOF2:POKEF,131:NEXT

560 FORF=F3TOF4:POKEF,176:NEXT

565 FORF=F1TOF4STEP64:POKEF,191:NEXT

570 FORF=F2TOF4STEP64:POKEF,191:NEXT

575 RETURN

600 REM ** INITIAL COMP-3 **

605 GOSUB650

610 X=RI+25:Y=RJ+25:GOSUB650

612 I=RI-3:J=RJ-3

615 C=0:D=0

620 RETURN

650 REM ** FETCH NEW-1 **

660 RI=RND(5):RJ=RND(5)

665 IFRI=3ANDRJ=3THEN660

670 RETURN

725 REM ** CON SENSE-1 **

730 FORXP=2TO2+ABS(I):FORYP=1TO1+ABS(J)

735 IFPOINT(X+SGN(I)*XP,Y+SGN(J)*YP)=-1THEN750

740 NEXT:NEXT

745 CO=0:RETURN

750 CO=1:RETURN
```

215

```
800 REM ** UD SCORE-1 **

805 U$="#.###"

810 E=D/C

845 PRINT@0,"DISTURB"

850 PRINT@15,"NO. CONTACTS"

852 PRINT@35,"NO. GOOD MOVES"

853 PRINT@55,"SCORE"

855 PRINT@87,C:PRINT@108,D

856 IFDC=0THEN860

857 PRINT@64,DT"OF 24"

860 PRINT@119,USINGU$;E

865 RETURN

1000 REM ** SHOW MEM-1 **

1010 W=16327

1015 FORM=1TO5:FORN=1TO5:FORO=1TO2

1020 POKEW,127+M(M,N,O)

1025 W=W+1:NEXT:NEXT:NEXT

1030 RETURN

1200 REM ** DISTURB-2 **

1205 DATA-2,-2,-2,-1,-2,0,-2,1,-2,2,-1,-2,-1,-1,-1,0,
     -1,1,-1,2,0,-2,0,-1,0,1, 0,2,1,-2,1,-1,1,0,1,1,1,
     2,2,-2,2,-1,2,0,2,1,2,2

1210 X=25:Y=25

1215 READI,J

1225 DC=1:C=0:D=0:RETURN
```
Beta Adaptation Compile Listing
```
10 REM ** BETA DISTURB COMPILE MASTER **

15 DIMM(5,5,2):DIMT(24,2)

20 TN=1:GOSUB500

25 GOSUB600

30 FORO=1TO2:FORN=1TO5:FORM=1TO5:M(M,N,O)=3:NEXTM,N,O

35 DT=0:DI=-2:DJ=-3
```

216

```
 40 CLS:GOSUB550

 45 X=X+I:Y=Y+J:GOSUB725

 50 IFCO=1THEN60

 55 GOTO45

 60 C=C+1:IF(DT=0ANDC>=25)OR(DT>0AND(C=50RC=25))
    THEN65ELSE70

 65 GOSUB900:IFC=0THEN110

 70 MI=I+3:MJ=J+3:RI=M(MI,MJ,1):RJ=M(MI,MJ,2)

 75 IFNOT(RI=3ANDRJ=3)THEN85

 80 GOSUB650:M(MI,MJ,1)=RI:M(MI,MJ,2)=RJ

 85 I=RI-3:J=RJ-3:GOSUB725

 90 IFCO=1THEN100

 95 D=D+1:GOSUB800:GOTO45

100 C=C+1:IF(DT=0ANDC>=25)OR(DT>0AND(C=50RC=25))
    THEN105ELSE80

105 GOSUB900:IFC<>0THEN80

110 DT=DT+1:IFDT<=24THEN40

115 TN=TN+1:IFTN<=CNTHEN25

120 GOSUB875

500 REM ** DIST HEAD-3 **

510 CLS:PRINTTAB(15)"RODNEY BETA DISTURB COMPILE"

517 INPUT"HOW MANY CREATURES";CN:RETURN

550 REM ** FIELD-1 **

552 F1=15553:F2=15615:F3=16193:F4=16255

555 FORF=F1TOF2:POKEF,131:NEXT

560 FORF=F3TOF4:POKEF,176:NEXT

565 FORF=F1TOF4STEP64:POKEF,191:NEXT

570 FORF=F2TOF4STEP64:POKEF,191:NEXT

575 RETURN

600 REM ** INITIAL COMP-3 **

610 GOSUB650:X=RI+25:Y=RJ+25

612 GOSUB650

615 I=RI-3:J=RJ-3:C=0:D=0
```

217

```
620  RETURN
650  REM ** FETCH NEW-1 **
660  RI=RND(5):RJ=RND(5)
665  IFRI=3ANDRJ=3THEN660
670  RETURN
725  REM ** CON SENSE-1 **
730  FORXP=2TO2+ABS(I):FORYP=1TO1+ABS(J)
735  IFPOINT(X+SGN(I)*XP,Y+SGN(J)*YP)=-1THEN750
740  NEXT:NEXT
745  CO=0:RETURN
750  CO=1:RETURN
800  REM ** UD DIST SCORE-1 **
805  U$="#.###"
810  E=D/C
815  PRINT@0,"CREAT.":PRINT@10,"PHASE"
850  PRINT@18,"NO. CONTACTS"
852  PRINT@35,"NO. GOOD MOVES"
853  PRINT@55,"SCORE"
855  PRINT@87,C:PRINT@108,D
857  PRINT@64,TN"OF"CN:PRINT@74,DT
860  PRINT@119,USINGU$;E
865  RETURN
875  REM ** DIST COMP DATA-1 **
877  TN=TN-1:U$="#.##"
880  DT=1
886  CLS:PRINTTAB(25)"LONG-TERM AVERAGES":PRINT"PHASE",
     "CONT 5","CONT 25"
890  PRINT:PRINTDT;:PRINTTAB(16)USINGU$;(T(DT,1)+TN)/
     (TN*5);:PRINTTAB(32)USING U$;(T(DT,2)+TN)/(TN*25)
892  INPUTS$:DT=DT+1:IFDT>24THEN880ELSE886
900  REM ** DIST COMP-1 **
915  IFDT=0THEN945
920  IFC=5THEN935
925  IFC=25THEN940
930  IFC<25RETURN
935  T(DT,1)=T(DT,1)+D:RETURN
940  T(DT,2)=T(DT,2)+D
945  GOSUB1200:RETURN
1200 REM ** DISTURB-2 **
1205 DJ=DJ+1:IFDJ=0ANDDI=0THEN1205
1210 IFDJ>2THEN1215ELSE1220
1215 DJ=-2:DI=DI+1
1220 I=DI:J=DJ:X=25:Y=25
1225 C=0:D=0:RETURN
```

218

Beta-II And
An Unexpected Discovery

Restructuring a BETA-I scheme to fit the definition of a Level-II mechanism isn't a very difficult task, especially if you use the ALPHA-II programs as a starting point. Just as you built a basic BETA-I program from the ALPHA-I version, you will use the ALPHA-II programs from Chapter 8 to build up the BETA-II system described in this chapter.

The idea is to give a BETA creature an extended sensory system, making it sensitive to the quality of different kinds of objects on the screen. The most important feature of the job is seeing how the Beta memory can be extended to include the additional sensory parameters.

The first program in this chapter is simple a conversion demonstration. It shows how you can use the flowchart for a standard BETA program and rework the operations to fit an entirely different programming scheme. Specifically, these BETA-II programs use the faster POKE and PEEK graphics in place of the slower SET and RESET versions used for all Level-I programs in this book.

The first program is listed as a BETA-II program, but it really does not include provisions for making it behave any differently from a BETA-I scheme. The potential for generating a BETA-II mode of behavior is inherent in the system, however.

The second program in this chapter lets the creature print tracks wherever it moves on the screen, You will find that the creature is insensitive to its own tracks, responding only to the border figure. This is true Level-II activity, because the creature is able to distinguish one of its own footprints from the border figure, and respond appropriately.

This program is actually the only proper sort of BETA trail-drawing scheme. You worked with a trail-drawing BETA-I program back in Chapter 10, but at that time, it was necessary to violate some of the most important definitions of real robotics in order to get the job done. This is a job for BETA-II, and you see how it all works out a bit later in this chapter, *without* violating any of the basic principles.

The third program in this chapter clearly demonstrates BETA-II learning and adaptive behavior. The screen is filled with obstacles of different kinds, and the BETA creature responds as though they are all the same sort of obstacle. The creature can sense the presence of the obstacles, but cannot distinguish them from the border figure.

It might seem that building this third program is a waste of time and effort. Why bother with a BETA-II creature that cannot tell the difference between one object and another? I originally devised the program because it represented the next logical step in the evolution of BETA-II adaptive intelligence. At the time, it seemed to be a rather routine matter brought about by logical neccesity rather than desire to make a big leap forward in the evolutionary scheme. And being a rather routine step, I thought I'd merely file away the results in my personal file of experiments and not bother to include it in this book. Something hapened a few days later, however, that changed my casual perspective on the matter. Nowadays, I am quite excited about the newly discovered significance of this second BETA-II program.

Before attempting to explain its special significance, I must describe the fourth program in this chapter. The fourth program, yet another logical step in the evolutionary scheme, allows the creature to sense the qualitative differences between the various objects on the screen.

The Beta memory is expanded so that the creature has a memory of how it dealt with the various objects in the past. The significant point is that the BETA-II creature can respond to one of the objects in a fashion that is quite different from the way it responds to the others. Each object on the screen plays a different role in the creature's activity. Exactly how the creature responds to a contact situation depends upon which object is involved in that particular contact.

Like BETA-I creatures, these BETA-IIs soon set up a certain habit pattern of motion. The obstacles included within the border figure play important roles in the configuration of each creature's basic habit pattern.

I expected that to happen, but little more than that. However, it didn't take long to find out that BETA-II creatures respond to disturbance operations in a fascinating (and wholly unexpected) way. Whenever a BETA-II is first allowed to establish its habit pattern of motion and then disturbed, it tends to go back to that same pattern—even at the same place on the screen!

It turns out that BETA-II creatures are goal-seeking creatures. I had not expected to observe any goal-seeking activity with anything less than a GAMMA-II. Let the critter extablish its habit pattern, move it somewhere else on the screen, and launch it with any motion code. It seeks out the original habit pattern and the same general place in the "room" it worked before.

Occasionally, the creature responds to a disturbance by setting up a new habit pattern somewhere else on the screen. And sometimes it works its way to the original position on the screen, but adds a couple of flourishes to the motion pattern. More often than not, however, the creature find its way to the original position on the screen and executes the very same habit pattern of motion.

The third program in this chapter became important *after* I noted the goal-seeking success of the creature built from the fourth program. Using the third program, the habit-pattern, goal-seeking features are present, but the creature isn't always so successful in carrying them out. In this case, a disturbance is just about as likely to create a new habit pattern at a new place on the screen.

The lesson to be learned from comparing the third and fourth programs in this chapter is that the ability to discern the differences between different objects, respond to each of them in a unique fashion, and remember all the responses, gives the creature an apparent appreciation for where it is located in the "room." The creature uses its knowledge of the different objects as cues to its location.

By way of comparison, a creature that is unable to distinguish the various objects from one another (as in the third program) cannot take positional cues from those objects, and thus owes no particular allegiance to playing in one special part of the screen.

If you have been schooled in traditional machine control technology, I think you will be able to appreciate the power and elegance of these notions. Just look around at other so-called "robots" capable of keeping track of their position in the environment. You run into things such as expensive stepping motors, elaborate pattern-recognition schemes, and, above all, very sophisticated logic, control, and memory systems.

All a BETA-II machine has to be able to do is sense contact with an object, determine the unique properties of that object, and then remember what action it took to get away from it. If you trick the machine by moving one of its cue objects, it will be upset temporarily, of course, But it will adapt to the situation in some way, attempting to set up a new habit pattern of motion or wandering around until it stumbles across the object you moved out of place.

Take an ordinary "robot" and disturb its environment in a similar way, and the thing gets all out of whack. The elaborate "room plan" stored in its memory no longer serves the proper purpose—at best, it will turn away from an object that no longer exists in its path. At worst, the mismatch between the original surroundings (stored in memory) and the new surroundings causes the machine to run completely out of sync with its environment. You then are the victim of those nasty little cartoons and comedy routines that portray robots as mindless critters which execute a certain activity entirely inappropriate to the situation at hand.

BETA-II CONVERSION DEMO

The practical significance of this BETA-II CONVERSION DEMO program is to set up a working base for loading the more elaborate BETA-II programs. The educational side of the matter is showing how an entirely different programming format can be adapted to an existing flowchart.

Flowchart for BETA-II CONVERSION DEMO

Figure 12-1 shows the flowchart for this program. Comparing it with the flowchart for ALPHA-II CONVERSION DEMO (Fig. 8-1), you will find the changes are quite consistent with those found by comparing flowcharts for ALPHA-I BASIC (Fig. 3-2) and BETA-I BASIC (Fig. 9-1). In short, the main difference between the ALPHA-II and BETA-II flowcharts are the memory operations required for the latter version.

The BETA-II process begins in the usual BETA-like fashion—dimensioning the Beta memory space, doing a short HEADING routine, initializing the Beta memory, drawing the border figure, and initializing the position and motion of the creature. Upon sensing contact with the border figure (remember this program does not include obstacles within the border figure), the program increments the NO. CONTACTS counter, saves the current values of I and J (the motion code), reads the contents of the Beta memory (using the current motion code as the address), and then checks to see if the situation has been dealt with in the past.

If creature has dealt with the situation in the past, the BEFORE conditional is satisfied and the system decodes the motion

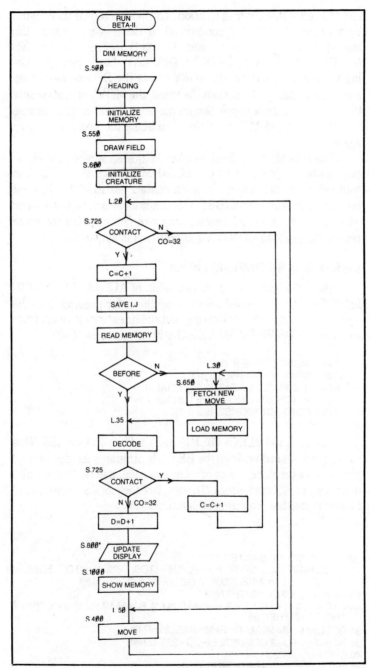

Fig. 12-1. Flowchart for BETA-II CONVERSION DEMO.

code stored at that address location. The motion code is then tested by a second CONTACT conditional. If there is no contact, the stored response is a good one; D=D+D increments the NO. GOOD MOVES figure, UPDATE DISPLAY shows the new scoring figures, and SHOW MEMORY prints the Beta memory map along the bottom of the screen. As usual, the MOVE operation sets the creature to its new position on the screen, and control returns to the first CONTACT conditional to begin the process all over again.

The only time the Beta memory is altered is when the situation leads the program to the LOAD MEMORY step. This can come about in two ways: finding a contact situation that has not been encountered before (BEFORE conditional is *not* satisfied) and when a previous stored contact response fails to work for some reason (the second CONTACT conditional is satisfied.)

Loading BETA-II CONVERSION DEMO

Since this program is an extension of ALPHA-II CONVERSION DEMO, it's a good idea to load the older program into the computer first. The following subroutines carry over from ALPHA-II CONVERSION DEMO with no changes at all:

```
S.400 MOVE (MOVE II-1)
S.550 DRAW FIELD (FIELD II-1)
S.600 INITIALIZE CREATURE (INITIAL II-1)
S.650 FETCH NEW MOVE (FETCH NEW II-1)
S.725 CONTACT (CONSENSE II-1)
```

Delete S.450 BLINK (BLINK II-1) at lines 450 and 455. This subroutine is omitted from the BETA-II programs for the sake of picking up a bit more program memory space, a move that is quite necessary for 4K computers. To complete the job, add or revise the following master program and subroutines:

```
10 REM ** BETA-II MASTER 1**
15 CLS: DIMM(5,5,2) :GOSUB50 :FORM=1TO5: FORN=1TO5 :FORO=1
   TO2:M(M,N, 0)=0:NEXTO,N,M:GOSUB550:GOSUB600
20 GOSUB725:IFCO=32THEN50
25 C=C+1: MI=I+3:MJ=J+3: RI=M(MI,MJ,1):RJ=M(MI,MJ,2) :IFNOT(RI=0
   AND RJ=0)THEN 35
30 GOSUB650 :M(MI,MJ,1)=RI:M(MI,MJ,2 =RJ
35 I=RI-3:J=RJ-3:GOSUB725:IFCO=32THEN45
40 C=C+1:GOTO30
45 D=D+1:GOSUB 800:GOSUB1000
50 GOSUB400:GOTO20
```

```
500 REM ** HEAD BII-1 **
505 CLS: PRINTTAB (25) "BETA-II":PRINTTAB (2Ø) "CONVERSION DE-
    MO":FORL.=1TO8:PRINT:NEXT:INPUT"DO 'ENTER' TO START";S$:
    CLS:RETURN
800 REM ** UD SCORE II-1 **
805 PRINT@ Ø, "NO. CONTACTS":PRINT@25,"NO GOOD MOVES"
    :PRINT 5Ø, "S CORE"
810 PRINT@7Ø,C:PRINT@95,D:PRINT@115,USING"#.##";D/C:RETURN
1000 REM ** SHOW MEM II-1 **
1005 W=16327:FORM=1TO5:FORN=1TO5:FORO=1TO2
1010 POKEW,127+M(M,N,O):W=W+1:NEXTO,N,M
1015 RETURN
```

It is a good idea to avoid the temptation of testing any part of
this program until it is entered in its entirety. The use of so many
POKE statements brings up the possibility of POKEing into the
program memory space and making a hopeless jumble of everything. Doublecheck your listing against the composite version at
the end of this chapter. Aside from looking for possible errors, look
for lines that can be deleted. When you are satisfied the program is
correct, save it on cassette tape *before* running it. Thus, if there are
any errors in the POKE statements, the fact that the program
becomes garbled will not make it necessary to enter the whole
thing from scratch again.

What to Expect From BETA-II CONVERSION DEMO

In most respects, this program runs just like BETA-I BASIC.
The creature works at the environment until it establishes a habit
pattern of motion. When this occurs, you will find that the memory
map no longer shows any changes and the SCORE figure rises
steadily toward perfection.

The most notable difference will be the higher speed of operation, an effect created by using the faster POKE graphics.

Since this is a rather basic and foundational program, there are
no provisions for disturbing the creature's activity. The only way to
break up an established habit pattern of motion is to strike the
BREAK key and run a whole new creature from the start.

BETA-II WITH TRAIL DRAWING

This program is a small variation of BETA-II CONVERSION
DEMO. It simply allows the creature to leave behind a set of
"footprints" wherever it goes on the screen. You can erase the
existing trail anytime you wish, however, and that makes it possible to view the habit pattern of motion quite clearly.

With the program for BETA-II CONVERSION DEMO in your
computer's program memory, make the following modifications to
the main program and subroutines S.4ØØ and S.5ØØ:

```
10 REM ** BETA-II MASTER 2 **
15 CLS:DIMM(5,5,2):GOSUB500FORM = 1 TO 5: FORN =1 TO 5: FORO
   =1TO2:M(M,N,O)=0:NEXTO,N,M:GOSUB550:GOSUB600
20 GOSUB725:IFCO=32ORCO=42THEN5
25 C=C+3:MJ=J+3:RI=M(MI,MJ,1):RJ=M(MI,MJ,2):IFNOT(RI= Ø AND
   RJ=0)THEN35

30 GOSUB650·M(MI,MJ,1)=RI:M(MI,MJ,2)=RJ
35 I=RI−3:J=RJ−3:GOSUB725:IFCO=32ORCO=42THEN45
40 C=C+1:GOTO30
45 D=D+1:GOSUB800:GOSUB1000
50 GOSUB40 :IFINKEY$Ø "T"THEN20
55 CLS:C=0:D=0:GOSUB550:GOTO20

400 REM ** MOVE BII-2 **
405 POKENP,42:NP=NP+I+64*J:POKENP,14Ø:RETURN

500 REM ** HEAD BII-2 **
505 CLS·PRINTTAB(25)"BETA-II":PRINTTAB(20)"TRAIL DRAWING VER-
SION
510 PRINT:PRINT"STRIKE THE 'T' KEY TO ERASE THE EXISTING TRAIL"
515 FORL=1TO4:PRINT:NEXT:INPUT"STRIKE THE 'ENTER' KEY TO
START";
S$:CLS:RETURN
```

Load the whole works onto cassette table before attempting to run the program. When you do run the program, you'll find the creature figure leaving behind an asterisk "footprint." This particular effect is created by the MOVE operation. Rather than erasing the creature figure before printing it in its new position, the POKE NP,42 statement in line 405 replaces the old creature figure with an asterisk. The little critter thus moves around its environment, leaving behind a trail of asterisks.

Of course, there is a very good chance that the trail will appear complicated and confusing through the first 20 contacts or so. But sooner or later the creature picks up a habit pattern of motion that carries it through the same path again and again.

To see the habit-pattern trail, without all the confusion generated during the initial learning phase, simply strike the *T* key. According to line 55 of the master program (BETA-II MASTER 2), this clears the screen, zeroes the score, redraws the border figure and returns operations back to line 20. Since the old set of tracks are eliminated along with everything else during the CLS operation, it figures that any new tracks will belong to the habit pattern of motion.

At the time of this writing, no one has yet generated a formal experimental study of BETA-II habit patterns. Someone ought to take the time to run this program any number of times, drawing and classifying the different kinds of habit patterns.

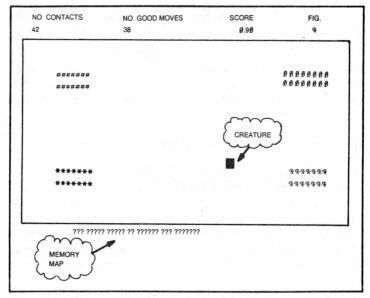

| NO. CONTACTS | NO. GOOD MOVES | SCORE | FIG. |
| 42 | 38 | 0.90 | % |

```
      #######                                    0 0 0 0 0 0 0
      #######                                    0 0 0 0 0 0 0


                                    CREATURE


      *******                                    % % % % % %
      *******                                    % % % % % %


      ??? ????? ????? ?? ?????? ??? ???????

         MEMORY
         MAP
```

Fig. 12-2. Flowchart for BETA-II WITH OBSTACLES AND MANUAL DIS-
TURB.

BETA-II WITH OBSTACLES AND MANUAL DISTURB

Now the fun begins. The screen format for this project, shown
in Fig. 12-2, includes a set of four obstacles within the border
figure. The creature's sensory capability is extended to include
sensing the qualitative differences between the four obstacles as
well as the top, bottom, and sides of the border figure itself.

The fact that the creature can sense the differences between
these objects is demonstrated by the FIG. heading at the upper
left-hand corner of the screen. As in the ALPHA-II experiments
you conducted earlier, the program displays the nature of the
obstruction in the creature's path. In Fig. 12-2, it so happens that
the creature made contact with the percent-sign obstacle, and you
see a % drawn under the FIG. heading on the display.

The creature for this program, however, is not equipped to
respond to the four extra obstacles in any way different from the
way it responds to the border figure. In a manner of speaking, the
creature thinks any one of the obstacles is simply some sort of
extension of the border figure. You will have a chance to change
that state of affairs in the last program in this chapter.

As mentioned earlier, I had not originally planned to include
this program in the book. It seemed to be a fairly trivial extension of

the BETA-II CONVERSION DEMO, a necessary logical step, but not a potentially interesting one. How wrong I was!

It turns out that this program serves as an important contrast to things you will observe in the last program in this chapter. So get it into your machine, play with it for a while, and save it for future reference. You will probably want to return to it again after finishing the last program for the BETA-II series.

Flowchart for BETA-II WITH OBSTACLES AND DISTURB

Figure 12-3 is the flowchart for this particular program. It differs from the BETA-II CONVERSION DEMO version in just two respects. First, note the S.675 DRAW FIGS subroutine now included in the start-up phase of the program. This subroutine is the one responsible for drawing the four figures which serve as extra obstacles for the creature. It ought to be a familiar subroutine because it was used in some of the ALPHA-II programs in Chapter 8.

The second difference is the DISTURB loop. Everytime the creature is moved to a new position at the MOVE operation, a DISTURB conditional checks to see whether or not you want to disturb the creature's activity. If not, operations return immediately to CONTACT. But, if you specify a disturbance by striking the *D* key, the DISTURB conditional leads to a series of three subroutines: DRAW FIELD, DRAW FIGS, and INITIALIZE CREATURE.

The disturbance loop actually begins with a CLS operation which clears the entire screen. The main reason for doing this is to remove the old creature's image. In any case, the border figure is redrawn, the figures (obstacles) are redrawn, and, most important to the purpose of the operation, the creature is set to a new initial position and given a randomly generated motion code.

The idea is to start running a new creature, give it time to establish a habit pattern of motion, and then disturb the pattern by striking the enter key. The disturbance operation does not affect the Beta memory in any way; when operations begin again, the creature already has access to some learned information about its surroundings.

You will find that the creature adapts to the disturbance in one of three ways: returning to the original habit pattern of motion, building up a new habit pattern, or returning to the original pattern at a different place or with some variations.

Loading BETA-II WITH OBSTACLES AND DISTURB

Load BETA-II CONVERSION DEMO into your machine and then make the following additions and modifications:

```
10 REM ** BETA-II MASTER 3 **
15 CLS:DIM(5,5,2):GOSUB50Ø:FORM=1TO5:FORN=1TO5: FORO=1TO2:
   M(M,N,Ø)=Ø:NEXTO,N,M:GOSUB55Ø:GOSUB675:GOSUB60Ø
20 GOSUB725:IFCO=32THEN5Ø
22 CI=CO
25 C=C+1:MI=I+3:MJ=J+3RI=M(MI,MJ,1):RJ=M(MI,MJ,2) IFNOT RI=Ø
   AND RJ=Ø THEN35
RJ=Ø)THEN35
30 GOSUB65Ø:M(MI,MJ,1)=RI:M(MI,MJ,2)=RJ
35 I=RI-3:J=RJ-3:GOSUB725:IFCO=32THEN45
40 C=C+1:GOTO3Ø
45 D=D+1:GOSUB80Ø:GOSUB10ØØ
50 GOSUB40Ø:IFINKEY$="D"THEN55ELSE2Ø
55 CLS:GOSUB55Ø:GOSUB675:GOSUB60Ø:GOTO2Ø
500 REM ** HEAD BII-3 **
505 CLS PRINTTAB(25)"BETA-II":PRINTTAB(15)"AVOID FIGURES AND
    DISTURB": FORL = 1TO8:PRINT :NEXT :INPUT "DO 'ENTER' TO
    START";S$:CLS:RETURN
800 REM ** UD SCORE II-2 **
805 PRINT @ Ø ,"NO. CONTACTS":PRINT @ 2Ø ,"NO. GOOD MOVES":
    PRINT 4Ø,"SCORE":  PRINT @ 6Ø ,"FIG."
810 PRINT @  65 C :PRINT @ 85 D :PRINT @1Ø5 USING "#. ##" ;D/C:
    PRINT @  125,CH R$(C1)
815 RETURN

600 REM ** INITIAL II-2 **
605 SF=Ø:NP=15968:GOSUB65Ø:I=RI-3:J=RJ-3
610 C= Ø:D=Ø:RETURN
```

Also add S.675 DRAW FIGS from Chapter 8. Doublecheck your listing against the composite listing at the end of this chapter, save the program on cassette tape, and then run it.

The more often you disturb this creature's activity, the more often it shows scores of 1.00. If you want to speed up the action a little bit, delete the GOSUB 1ØØØ statement in line 45 of the master program. You'll no longer see the memory map, but the deletion eliminates the relatively long period of time required for updating it.

BETA-II WITH INDEPENDENT RESPONSES AND DISTURB

This is the program that surprised me. It clearly shows goal-seeking behavior for a system I thought too simple for that sort of

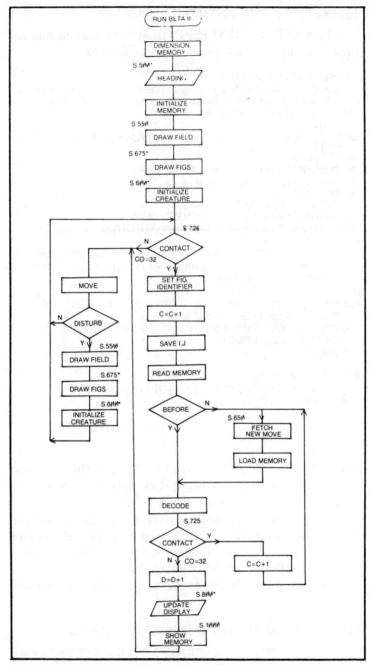

Fig. 12-3. Flowchart for BETA-II WITH OBSTACLES AND DISTURB.

activity. The program is an extension of the previous one. In this case, three of the obstacles on the screen are treated as something quite different from the other obstacles and the border figure. The family of possible responses is the same, but the creature has an extended memory that allows different sets of responses in each case.

The original purpose of the project was to show how the Beta memory has to be expanded to cover the idea of having different responses to different objects. You will see in the master program the Beta memory is expanded to a 5×5×2×2, as opposed to the smaller 5×5×2 for earlier BETA-II programs.

It would be nice if the creature could make an independent set of responses to all four different obstacles contained within the border figure. However, that calls for a 5×5×4×2 memory dimension which exceeds the limitations of a 4K memory system. So for the sake of experimenters having smaller memory systems, the obstacles built from Øs and %s are treated as part of the border figure. Even then, you must do a CLEAR 16 before the program can be run on a 4K system.

You will be able to see the creature responding to the border, #, and * figures with different motions, even though the contact situations are otherwise identical.

The unexpected, and now most important, feature of this scheme is the way the creature uses the obstacles as cues to its position on the screen. Nearly every time you disturb this creature by striking the *D* key, you will find it working its way back to the original screen position and habit pattern.

Sometimes the creature works out a new position and habit pattern, and sometimes the pattern is in the same position on the screen, but with a few interesting variations. Nevertheless, one gets the impression the creature is motivated to seek out a particular position and pattern of motion. Considering you have not programmed the machine to do that kind of task, you can be sure you are seeing the seed of genuine machine adaptive intelligence at work.

Flowchart for BETA-II With Independent Responses and Disturb

The flowchart in Fig. 12-4 is quite similar to the one featured in the previous section of this chapter. Notable differences are the absense of a HEADING operation, an INITIALIZE MEMORY step, and a SHOW MEMORY operation. These steps have been eliminated so that the program fits into a 4K-computer format.

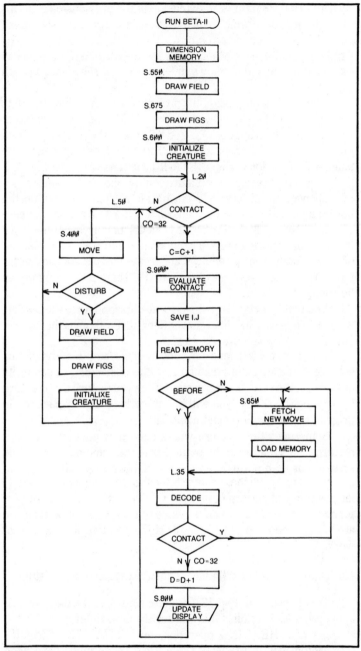

Fig. 12-4. Flowchart for BETA-II WITH INDEPENDENT RESPONSES AND DISTURB.

You probably have realized by now that the HEADING operation in most of the programs is not really essential. It merely serves as a convenient way to identify the program residing in the system. SHOW MEMORY is eliminated here for a couple of reasons: first, to allow more room for Beta memory, and second because it is difficult to show the entire memory anyway. This program runs with 5×5×2×2, or 100 Beta memory elements, as opposed to 50 for earlier BETAs. SHOW MEMORY isn't critical to the operation.

INITIALIZE MEMORY can be omitted because most computer systems automatically initialize all memory elements to zero upon executing the RUN command. So why bother with a special INITIALIZE MEMORY operation when, in this case, you would want the memory initialized with Øs anyway?

One important addition to the flowchart is the EVALUATE CONTACT operation. This step is responsible for sensing the nature of the contact figure and setting a third element in the Beta memory address. You see, the memory is not only addressed by the I and J values of the motion code (as with all earlier BETA programs), but the quality of the obstacle as well.

So now there are three distinct elements in the memory address: the current I value, the current J value, and a number representing the nature of the obstacle: border, asterisk, or pound sign. All other operations, including the DISTURBANCE phase, are the same as those used with BETA-II WITH OBSTACLES AND DISTURB.

Loading BETA-II With Independent Responses and Disturb

This program is fairly easy to load if BETA-II WITH OBSTA-CLES AND DISTURB is already resident in program memory. You have to delete some unnecessary subroutines, modify the master program, and add just one new subroutine. Delete these from the older program:

S.500 HEADING (lines 5ØØ-5Ø5)
S.1000 SHOW MEMORY (lines 1ØØØ-1Ø15).

Then modify or add these:

```
10 REM ** BETA-II MASTER 4 **
15 CLS:DIMM(5,5,2,2): GOSUB 550:GOSUB675:GOSUB6ØØ
20 GOSUB725IFCO= 32THEN5Ø
22 C1=CØ
25 C=C+1:GOSUB9ØØ:MI=1+3:MJ=J+3:RI=M (MI,MJ,B,1):
     RJ=M(MI,MJ,B,2): IFNOT(RI=Ø ANDRJ=Ø)THEN 35
30 GOSUB65Ø:M(MI,MJ,B,1)=RI:M(MI,MJ,B,2)=RJ
35 I=RI-3:J=RJ-3:GOSUB725:IFCO=32THEN45
```

```
40 C=C+1:GOTO30
45 D=D+1:GOSUB800
50 GOSUB400:IFINKEY$="D"THEN55ELSE 20
55 CLS:GOSUB550:GOSUB675:GOSUB600:GOTO20
900 REM ** CON EVAL BII-1 **
905 IFCO=131ORCO=176ORCO=191THEN925
910 IFCO=-42THEN930
915 IFCO=35THEN935
925 B=0:RETURN
930 B=1:RETURN
935 B=2:RETURN
```

Doublecheck your entire listing against the composite version at
the end of this chapter. Load the program onto cassette tape, enter
a CLEAR 16, and you're ready to go.

The CLEAR 16 operation is absolutely necessary for computers having less than 16K memories. Doing the CLEAR 16 shifts
most of the normal memory space allocated for string variables to
usable program space. Without doing the CLEAR 16 before running this program, small computers will show an out-of-memory
error.

There is no HEADING routine, so the program begins shortly
after you enter the RUN command. Watch the creature gradually
establi h a habit pattern of motion, then strike the *D* key to upset it.
In most instances you will see the creature working its way back to
the same position and habit pattern established earlier. Talented
little bugger, isn't it?

COMPOSITE PROGRAM LISTINGS

BETA-II CONVERSION DEMO Listing

```
10 REM ** BETA-ii MASTER 1 **
15 CLS:DIMM(5,5,2):GOSUB500:FORM=1TO5:FORN=1TO5:FORO
   =1TO2:M(M,N,O)=0:NEXTO,N, M:GOSUB550:GOSUB600

20 GOSUB725:IFCO=32THEN50

25 C=C+1:MI=I+3:MJ=J+3:RI=M(MI,MJ,1):RJ=M(MI,MJ,2):
   IFNOT(RI=0ANDRJ=0)THEN35

30 GOSUB650:M(MI,MJ,1)=RI:M(MI,MJ,2)=RJ

35 I=RI-3:J=RJ-3:GOSUB725:IFCO=32THEN45

40 C=C+1:GOTO30

45 D=D+1:GOSUB800:GOSUB1000

50 GOSUB400:GOTO20

400 REM ** MOVE II-1 **
```

```
405 POKENP,32:NP=NP+I+64*J:POKENP,140:RETURN

450 REM ** BLINK II-1 **

455 POKENP,32:POKENP,140:RETURN

500 REM ** HEAD BII-1 **

505 CLS:PRINTTAB(25)"BETA-II":PRINTTAB(20)"CONVERSION
    DEMO":FORL=1TO8:PRINT:N
EXT:INPUT"DO 'ENTER' TO START";S$:CLS:RETURN

550 REM ** FIELD II-1 **

555 FORF=15553TO15615:POKEF,131:NEXT

560 FORF=16193TO16255:POKEF,176:NEXT

565 FORF=15553TO16255STEP64:POKEF,191:NEXT

570 FORF=15615TO16255STEP64:POKEF,191:NEXT

575 RETURN

600 REM ** INITIAL II-1 **

605 SF=0:NP=15968:GOSUB650:I=RI-3:J=RJ-3:RETURN

650 REM ** FETCH NEW II-1 **

655 RI=RND(5):RJ=RND(5):IFSF=0ANDRI=3ANDRJ=3THEN655

660 RETURN

725 REM ** CON SENSE II-1 **

730 FORSI=ABS(I)TO1STEP-1:FORSJ=ABS(J)TO1STEP-1

735 CO=PEEK(NP+SI*SGN(I)+64*SJ*SGN(J))

740 IFCO<>32RETURN

745 NEXTSJ,SI:RETURN

800 REM ** UD SCORE II-1 **

805 PRINT@0,"NO. CONTACTS":PRINT@25,"NO. GOOD MOVES":
    PRINT@50,"SCORE"

810 PRINT@70,C:PRINT@95,D:PRINT@115,USING"#.##";D/C:RETURN

1000 REM ** SHOW MEM II-1 **

1005 W=16327:FORM=1TO5:FORN=1TO5:FORO=1TO2

1010 POKEW,127+M(M,N,O):W=W+1:NEXTO,N,M

1015 RETURN
```

BETA-II WITH TRAIL DRAWING Listing

```
10 REM ** BETA-II MASTER 2 **

15 CLS:DIMM(5,5,2):GOSUB500:FORM=1TO5:FORN=1TO5:FORO
   =1TO2:M(M,N,O)=0:NEXTO,N,M:GOSUB550:GOSUB600
```

```
20 GOSUB725:IFCO=320RCO=42THEN50

25 C=C+1:MI=I+3:MJ=J+3:RI=M(MI,MJ,1):RJ=M(MI,MJ,2)
   :IFNOT(RI=0ANDRJ=0)THEN35

30 GOSUB650:M(MI,MJ,1)=RI:M(MI,MJ,2)=RJ

35 I=RI-3:J=RJ-3:GOSUB725:IFCO=320RCO=42THEN45

40 C=C+1:GOTO30

45 D=D+1:GOSUB800:GOSUB1000

50 GOSUB400:IFINKEY$<>"T"THEN20

55 CLS:C=0:D=0:GOSUB550:GOTO20

400 REM ** MOVE BII-2 **

405 POKENP,42:NP=NP+I+64*J:POKENP,140:RETURN

500 REM ** HEAD BII-2 **

505 CLS:PRINTTAB(25)"BETA-II":PRINTTAB(20)"
    TRAIL-DRAWING VERSION

510 PRINT:PRINT"STRIKE THE 'T'
    KEY TO ERASE THE EXISTING TRAIL"

515 FORL=1TO4:PRINT:NEXT:INPUT"STRIKE THE
    'ENTER' KEY TO START";S$:CLS:RETURN

550 REM ** FIELD II-1 **

555 FORF=15553TO15615:POKEF,131:NEXT

560 FORF=16193TO16255:POKEF,176:NEXT

565 FORF=15553TO16255STEP64:POKEF,191:NEXT

570 FORF=15615TO16255STEP64:POKEF,191:NEXT

575 RETURN

600 REM ** INITIAL II-1 **

605 SF=0:NP=15968:GOSUB650:I=RI-3:J=RJ-3:RETURN

650 REM ** FETCH NEW II-1 **

655 RI=RND(5):RJ=RND(5):IFSF=0ANDRI=3ANDRJ=3THEN655

660 RETURN

725 REM ** CON SENSE II-1 **

730 FORSI=ABS(I)TO1STEP-1:FORSJ=ABS(J)TO1STEP-1

735 CO=PEEK(NP+SI*SGN(I)+64*SJ*SGN(J))

740 IFCO<>32RETURN
```

```
745 NEXTSJ,SI:RETURN

800 REM ** UD SCORE II-1 **

805 PRINT@0,"NO. CONTACTS":PRINT@25,"NO. GOOD MOVES":
    PRINT@50,"SCORE"

810 PRINT@70,C:PRINT@95,D:PRINT@115,USING"#.##";D/C:RETURN

1000 REM ** SHOW MEM II-1 **

1005 W=16327:FORM=1TO5:FORN=1TO5:FORO=1TO2

1010 POKEW,127+M(M,N,O):W=W+1:NEXTO,N,M

1015 RETURN
```

BETA-II WITH OBSTACLES AND MANUAL DISTURB Listing

```
10 REM ** BETA-II MASTER 3 **

15 CLS.DIMM(5,5,2):GOSUB500:FORM=1TO5:FORN=1TO5
   :FORO=1TO2:M(M,N,O)=0:NEXTO,N,
   M:GOSUB550:GOSUB675:GOSUB600

20 GOSUB725:IFCO=32THEN50

22 C1=CO

25 C=C+1:MI=I+3:MJ=J+3:RI=M(MI,MJ,1):RJ=M(MI,MJ,2)
   :IFNOT(RI=0ANDRJ=0)THEN35

30 GOSUB650:M(MI,MJ,1)=RI:M(MI,MJ,2)=RJ

35 I=RI-3:J=RJ-3:GOSUB725:IFCO=32THEN45

40 C=C+1:GOTO30

45 D=D+1:GOSUB800:GOSUB1000

50 GOSUB400:IFINKEY$="D"THEN55ELSE20

55 CLS:GOSUB550:GOSUB675:GOSUB600:GOTO20

400 REM ** MOVE II-1 **

405 POKENP,32:NP=NP+I+64*J:POKENP,140:RETURN

450 REM ** BLINK II-1 **

455 POKENP,32:POKENP,140:RETURN

500 REM ** HEAD BII-3 **

505 CLS:PRINTTAB(25)"BETA-II":PRINTTAB(15)
    "AVOID FIGURES AND DISTURB":FORL=1T
    O8:PRINT:NEXT:INPUT"DO 'ENTER' TO START";S$:CLS:RETURN

550 REM ** FIELD II-1 **
```

237

```
555 FORF=15553T015615:POKEF,131:NEXT

560 FORF=16193T016255:POKEF,176:NEXT

565 FORF=15553T016255STEP64:POKEF,191:NEXT

570 FORF=15615T016255STEP64:POKEF,191:NEXT

575 RETURN

600 REM ** INITIAL II-2 **

605 SF=0:NP=15968:GOSUB650:I=RI-3:J=RJ-3

610 C=0:D=0:RETURN

650 REM ** FETCH NEW II-1 **

655 RI=RND(5):RJ=RND(5):IFSF=0ANDRI=3ANDRJ=3THEN655

660 RETURN

675 REM ** FIGS II-1 **

677 P=15755

680 FORF=0TO1:FORN=P+64*FTOP+64*F+8:POKEN,35:NEXTN,F

682 FORF=0TO1:FORN=P+32+64*FTOP+40+64*F:POKEN,48:NEXTN,F

684 FORF=0TO1:FORN=P+192+64*FTOP+200+64*F:POKEN,42:NEXTN,F

686 FORF=0TO1:FORN=P+224+64*FTOP+232+64*F:POKEN,37:NEXTN,F

690 RETURN

725 REM ** CON SENSE II-1 **

730 FORSI=ABS(I)TO1STEP-1:FORSJ=ABS(J)TO1STEP-1

735 CO=PEEK(NP+SI*SGN(I)+64*SJ*SGN(J))

740 IFCO<>32RETURN

745 NEXTSJ,SI:RETURN

800 REM ** UD SCORE II-2 **

805 PRINT@0,"NO. CONTACTS":PRINT@20,"NO. GOOD MOVES"
    :PRINT@40,"SCORE":PRINT@6 0,"FIG."

810 PRINT@65,C:PRINT@85,D:PRINT@105,USING"#.##";D/C:PRINT
    @125,CHR$(C1)

815 RETURN

1000 REM ** SHOW MEM II-1 **

1005 W=16327:FORM=1TO5:FORN=1TO5:FORO=1TO2

1010 POKEW,127+M(M,N,O):W=W+1:NEXTO,N,M

1015 RETURN
```

238

BETA-II WITH INDEPENDENT RESPONSE AND DISTURB Listing

```
10 REM ** BETA-ii MASTER 4 **

15 CLS:DIMM(5,5,2,2):GOSUB550:GOSUB675:GOSUB600

20 GOSUB725:IFCO=32THEN50

22 C1=CO

25 C=C+1:GOSUB900:MI=I+3:MJ=J+3:RI=M(MI,MJ,B,1):RJ=M
   (MI,MJ,B,2):IFNOT(RI=0AND
RJ=0)THEN35

30 GOSUB650:M(MI,MJ,B,1)=RI:M(MI,MJ,B,2)=RJ

35 I=RI-3:J=RJ-3:GOSUB725:IFCO=32THEN45

40 C=C+1:GOTO30

45 D=D+1:GOSUB800

50 GOSUB400:IFINKEY$="D"THEN55ELSE20

55 CLS:GOSUB550:GOSUB675:GOSUB600:GOTO20

400 REM ** MOVE II-1 **

405 POKENP,32:NP=NP+I+64*J:POKENP,140:RETURN

450 REM ** BLINK II-1 **

455 POKENP,32:POKENP,140:RETURN

550 REM ** FIELD II-1 **

555 FORF=15553TO15615:POKEF,131:NEXT

560 FORF=16193TO16255:POKEF,176:NEXT

565 FORF=15553TO16255STEP64:POKEF,191:NEXT

570 FORF=15615TO16255STEP64:POKEF,191:NEXT

575 RETURN

600 REM ** INITIAL II-2 **

605 SF=0:NP=15968:GOSUB650:I=RI-3:J=RJ-3

610 C=0:D=0:RETURN

650 REM ** FETCH NEW II-1 **

655 RI=RND(5):RJ=RND(5):IFSF=0ANDRI=3ANDRJ=3THEN655

660 RETURN

675 REM ** FIGS II-1 **

677 P=15755
```

239

```
680 FORF=0T01:FORN=P+64*FTOP+64*F+8:POKEN,35:NEXTN,F

682 FORF=0T01:FORN=P+32+64*FTOP+40+64*F:POKEN,48:NEXTN,F

684 FORF=0T01:FORN=P+192+64*FTOP+200+64*F:POKEN,42:NEXTN,F

686 FORF=0T01:FORN=P+224+64*FTOP+232+64*F:POKEN,37:NEXTN,F

690 RETURN

725 REM ** CON SENSE II-1 **

730 FORSI=ABS(I)TO1STEP-1:FORSJ=ABS(J)TO1STEP-1

735 CO=PEEK(NP+SI*SGN(I)+64*SJ*SGN(J))

740 IFCO<>32RETURN

745 NEXTSJ,SI:RETURN

800 REM ** UD SCORE II-2 **

805 PRINT@0,"NO. CONTACTS":PRINT@20,"NO. GOOD MOVES":
    PRINT@40,"SCORE":PRINT@6 0,"FIG."

810 PRINT@65,C:PRINT@85,D:PRINT@105,USING"#.##";D/C:PR
    INT@125,CHR$(C1)

815 RETURN

900 REM ** CON EVAL BII-1 **

905 IFCO=131ORCO=176ORCO=191THEN925

910 IFCO=42THEN930

915 IFCO=35THEN935

925 B=0:RETURN

930 B=1:RETURN

935 B=2:RETURN
```

240

13

A Return To BETA-I:
With Confidence This Time

The BETA-II concepts presented in the previous chapter offer some real challenges in the way of original experimental work and, of course, some possibilities for nifty robot games. For our immediate purposes, it is time to get back onto the main evolutionary track. The material in this chapter marks a return to the basic BETA-I concept—an Alpha-Class machine outfitted with a Beta-Class memory. You will find an entirely new feature injected into the scheme, and this new feature paves the way for the next big evolutionary jump to Gamma-Class behavior.

BETA WITH CONFIDENCE

Suppose you are running an ordinary BETA-I program, not necessarily a fancy compiling program, but just one of the basic versions with scoring and memory showing. Recall that such a creature sooner or later locks into a habit pattern of motion. The SCORE figure rises steadily, the little memory map along the bottom of the screen no longer shows any changes, and you might even be able to trace the habit pattern as the BETA creature moves around on the screen.

Now suppose the creature is not only remembering responses that work, but also counts the number of times each response works. Sounds simple enough, doesn't it? Well if that's the case, it figures that the responses taking part in the habit pattern of motion are going to pick up some pretty large counts. Every time the creature completes its excursion through a habit pattern, each of the responses involved are counted or incremented, by one.

In a manner of speaking, the creature picks up some confidence regarding the responses that work for it time and time again. This confidence level is expressed by the counters that keep track of the number of times a given response works. The more often a certain response, or set of responses, works, the more confident the creature becomes.

But suppose something happens to upset the creature's habit pattern of motion. The confidence counters have accumulated some large numbers because the responses worked so many times before. The conditions are upset now, though, and suddenly one or more of those "confident" responses don't work right. What should the creature do? How about having it execute that confident response again? If the response doesn't work, it tries it again, and again, and again.

Every time the creature tries a "confident" response and it doesn't work, the confidence counter is decremented. The little guy, you see, is losing confidence now. And when the confidence counter for that particular response gets down to zero, the appropriate action is to go after an Alpha-like random response. Eventually the creature will pick a response that works, assign some minimum confidence level to it, and go on from there. Every time this new response works, the confidence level picks up.

It takes between 20 and 30 contacts, on the average, for a BETA creature to establish a high-confidence habit pattern of motion. Once it gets there, it takes some work to find out that it no longer works. The BETA creature with confidence must work at breaking an old habit and establishing a new one.

In the next program you are going to be building an ordinary BETA-I creature, with a scheme inserted for counting the number of times it deals with a given situation, both effectively and non-effectively. Each time a response works, it is stored in memory along with a number representing the number of times that response works. The confidence numbers can go as high as four. (There seems to be little gained by having them go any higher). Each time a response does not work, the confidence level associated with it is decreased by one, down as far as zero.

On a simple BETA-I level, the effect of the confidence level feature isn't noticeable at first. In fact, if the creature adapts a simple habit pattern of motion one that is never disrupted in any significant way, the confidence effect will never show up. But, when something happens to shake the creature's confidence in previously successful responses, you'll find it banging its head against the wall a number of times before it finally gives up and tries something new.

A BETA-I creature that does not have a confidence feature owes no allegiance to a response that worked many times before. It might have solved a particular problem in one particular way a hundred times before, but the first time that same response doesn't work, the critter is willing to chuck all past experience and try something entirely new.

BETA with confidence is simply more stubborn. If a response to a particular situation worked before, the creature figures it will work again. It insists it will work again until continued effort finally shakes the confidence level down to zero, then it tries something else. In the meantime, its SCORE figure plunges steadily downward.

BETA creatures with confidence tend to show lower scoring figures than creatures without the confidence feature. Both score equally well until something happens to upset previously successful responses, and both recover from the disturbance at about the same rate. The scoring for a BETA with confidence plunges much lower when its faith is shaken. Finding that a previously good response no longer works is far more traumatic for a BETA with confidence, and the scores reflect that fact.

So why design a machine that seems to insist a certain response will work when, indeed, it no longer does? Why build in a feature that causes traumatic responses that aren't really necessary for accomplishing the task at hand? Those are good questions when they come from one trained in the traditional doctrines of engineering and control technology. To become a successful roboticist, however, one must be willing and able to set aside conventional doctrines and think in much broader, more general terms. The objective here is not that of building a machine to solve problems in the fastest and most efficient manner, but rather to give the machine a potential for solving problems no one thought it would ever encounter. To that end, there must be some trade offs.

The trade off in this case is something you might want to call adaptive trauma. This matter of adaptive trauma is extremely inefficient and, in fact, counterproductive; it tends to work against the machine's success.

Actually, there is a very simple and rational reason for introducing the confidence-level scheme and allowing adaptive trauma to mess up the creatures' scores. The reason is this: The feature is absolutely necessary for Gamma-Class behavior. So why deal with it in the context of a Beta-Class machine? There are two good answers to the last question.

First, it is tough enough to learn the fundamentals of Gamma-Class behavior without trying to see how the confidence-level feature works at the same time. It will be easier to appreciate the workings of a GAMMA-I creature if you already understand confidence levels in the context of the more familiar and simpler BETA system.

The second reason for dealing with confidence levels in a BETA context is that you must have some new scoring curves for reference purposes. The confidence-level feature is an indispensable part of GAMMA operation, and since the use of confidence levels tends to lower scoring curves, a direct comparison between GAMMA adaptive curves and those you have generated from non-confident BETAs would not be a valid one.

A SIMPLE BETA-I WITH CONFIDENCE

The introduction of a confidence-level subroutine and its ultimate function in Gamma-Class operations makes it necessary to begin tightening up the memory space devoted to master programs and their associated subroutines. Thus, you will find the programs and subroutines appearing in a much tighter and, alas, less flexible

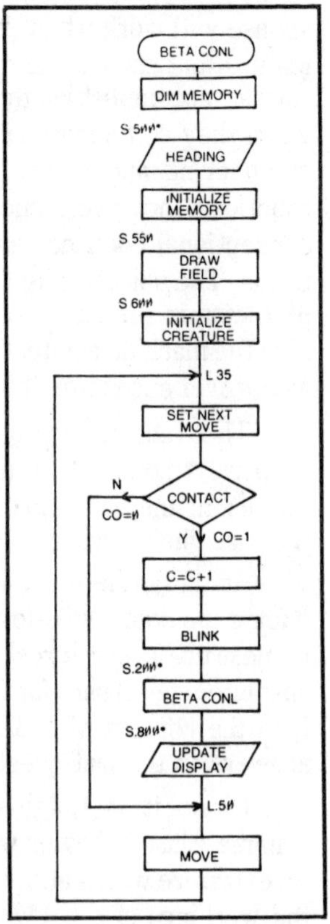

Fig. 13-1. Master program flowchart for BETA-I WITH CONFIDENCE.

form. There will be a lot more multiple-statement lines and, in some instance, fewer separate subroutines.

For example, look at the flowchart for the master program used with this BETA-I WITH CONFIDENCE creature. It appears rather simple compared to the master programs you have been using lately. Don't be mislead by the simplicity of the drawing, however. The fact of the matter is that much of the usual ALPHA and BETA operations have been shifted over to a subroutine represented by BETA CONL in Fig. 13-1.

The flowchart in Fig. 13-1 shows the operation starting out much as an ordinary BETA-I program does. The memory is dimensioned, a heading appears on the screen, the field is drawn, and the creature is initialized. There's nothing significantly new so far. Then, the flowchart calls for a SET NEW MOVE, followed immediately by a typical CONTACT conditional. If there is no contact, $CO=\emptyset$ and operation shifts down to the MOVE step. And from there, the creature sets its next move. If any suggested move results in a contact situation, the CONTACT conditional is satisfied, the NO. CONTACTS figure is incremented by $C=C+1$, the creature figure is flashed by BLINK, and then the system calls BETA CONL—the confidence-level subroutine. The important point in this part of the discussion is that the confidence-level feature is put into effect only when the creature is dealing with a contact situation.

Now, look at the flowchart in Fig. 13-2. It is the flowchart for the BETA CONL operation, $S.2\emptyset\emptyset$ BETA CONL-1. Many of the operations on this flowchart should look familiar; they have been part of BETA work you have been doing through the past few chapters.

In the first step, for instance, the current values of I and U (the motion code) are saved. The second step calls for addressing the Beta memory and then the system encounters the BEFORE conditional. It turns out that the BEFORE conditional is defined much differently than before. Instead of looking for a stop code stored in memory ($I=\emptyset,J=\emptyset$, this version looks at the confidence level, KL, in the memory data. If KL is equal to zero, the implication is that the creature hasn't the foggiest idea how to solve the problem at hand, and as a result, the system shifts over to an Alpha-like series of operations: BLINK, FETCH NEW MOVE, etc.

If BEFORE shows that the confidence level of the situation is greater than zero, it follows that a workable response from the past exists in the Beta memory. The system reads that motion code at

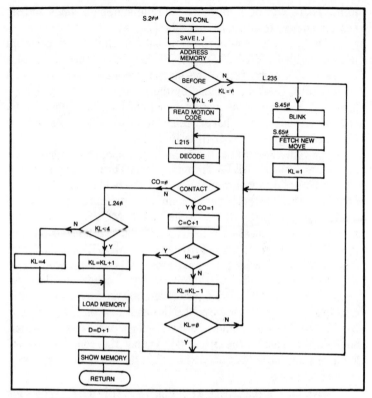

Fig. 13-2. Flowchart for subroutine 2ØØ, BETA CONL.

READ MOTION CODE, decodes it to a stnadard I/J format, and tries it by means of a CONTACT conditional.

Now, if the response is still a good one, the CONTACT conditional is not satisfied and the system looks at another conditional, KL *less than* 4. In other words, "Is the confidence level less than the maximum?" If so, the confidence level is incremented by KL=KL+1. The previously established response to the contact situation worked again (CONTACT was not satisfied), and it is time to upgrade the creature's confidence level a little bit. There isn't much point in working with confidence levels greater than 4, however. So if it turns out that KL *less than* 4 is not satisfied, the KL=4 simply holds the confidence level at 4.

How does the system know what the KL value is? It read the value from memory back at the BEFORE conditional. It had to read the KL value at that time in order to determine whether or not the creature had ever encountered and solved the situation before.

246

Assuming the remembered response worked, the memory is revised at LOAD MEMORY to include the new KL value. Then the NO. GOOD MOVES figure is incremented at D=D+1 and the memory map is updated by SHOW MEMORY. The whole CONL subroutine is terminated and operations return to the main program.

The discussion thus far assumes (1) the creature has encountered a given situation before and remembers the workable solution and (2) the remembered solution works again. But what happens if the BEFORE conditional finds KL=∅?

If KL=∅, one of two things has happened. Either the situation at hand is an entirely new one as far as the creature is concerned, or a high-confidence response from the past has been tried and proven no good at all. In the first case, KL has never been incremented above zero, and in the second case, some higher value of KL has been decremented to zero. In either case, the creature has to pick a random response to get things going again.

This scheme uses the familiar S.65∅ FETCH NEW MOVE subroutine for picking out a random response. The difference in this confidence-level scheme is that KL is set to one, the lowest confidence level that means anything.

After the confidence level is set to one, the new "idea" is decoded and tested. If it doesn't work (as determined by the CONTACT conditional, the NO. CONTACTS counter is incremented and KL is tested for a zero value. Of course, KL is not equal to zero now because it was previously set to one by KL=1. The response didn't work, however, so the weak confidence level is decremented to zero by KL=KL−1. And, since KL=∅ again in this example, the system resorts to picking another random motion code.

Things continue running through this loop until the creature comes up with a motion code that solves the situation. The CONTACT conditional is not satisfied, KL is incremented to two by KL=KL+1, and operations soon get back to normal.

The only other confidence-level situation to consider is one where KL is greater than one, but for some reason the stored response doesn't work. Recall that one of the primary characteristics of BETA creatures is their ability to optimize the responses stored in Beta memory. It is a sad fact of life that some responses perfectly workable under one set of conditions simply do not work under slightly different conditions. BETAs cover this sort of situation by testing the validity of each remembered response, changing them whenever the situation demands it.

You have read about this in an earlier chapter, but perhaps a specific example would be helpful. The drawing in Fig. 13-3 shows the path taken by a BETA creature that has established a certain kind of habit pattern of motion. The pattern is one where the critter moves downward and a bit to the right. Upon making a contact with the border figure, it has learned to respond by moving upward and a bit less to the right. After that, it responds to a contact situation by moving downward and toward the right again. This regular pattern works out rather well until the creature hits the right-hand edge of the border figure. The learned response, moving upward and a bit toward the right, no longer works. The response that worked so many times before does not work under all possible situations of the kind, and the critter is forced to come up with a different kind of response. The new response will then (most likely) work, whether the contact is with the bottom of the border or with the right-hand edge.

This situation is dealt with on the flowchart in Fig. 13-2 at the point where the first KL=Ø conditional, the one immediately following C=C+1, is not satisfied. A response that worked before (BEFORE conditional satisfied) no longer works (CONTACT conditional satisfied.) As a result, the confidence level is decremented by one (KL=KL−1), and if the confidence level is still greater than zero, the stubborn critter tries the same response again. Note the loop back to DECODE.

This stubborn creature continues working the same old response until KL is decremented all the way down to zero. When

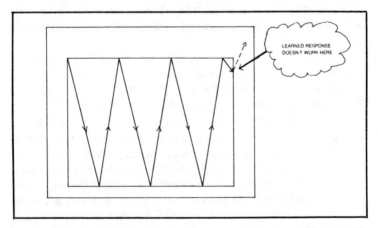

Fig. 13-3. An example showing how a previously learned response will not work under certain circumstances.

that finally happens, he wises up and fetches a random response by going to the loop beginning at L. 235.

Incidentally, this isn't altogether a matter of being stupid or mule-headed. If you were working with a genuine mechanical robot in the real world (as opposed to a simulated robot under ideal conditions), the fact that it tries the same response several times in succession might well get it out of the contact situation. Think about that for a moment.

Loading BETA-I WITH CONFIDENCE

The program can be loaded beginning with BETA-I BASIC WITH SCORING AND SHOWING MEMORY (Chapt. 9.) The following subroutines carry over to this new program without any necessary alterations:

```
S.450 BLINK or MOVE (BLINK-1)
S.550 DRAW FIELD (FIELD-2)
S.600 INITIALIZE CREATURE (INITIAL-2)
S.650 FETCH NEW MOVE (FETCH NEW-1)
S.725 CONTACT (CON SENSE-1)
S.1000 SHOW MEMORY (SHOW MEM-1)
```

Here is the S.$5\emptyset\emptyset$ HEADING subroutine, somewhat tightened up and shortened to eliminate the memory-gobbling FOR-NEXT loop:

```
500 REM ** BETA CONL HEAD-1 **
505 CLS:PRINTTAB(25) "RODNEY BETA-1":PRINTTAB(25)"CONL VER-SION"
510 INPUT"DO ENTER TO START";S$:RETURN
```

When comparing S.$2\emptyset\emptyset$ BETA CONL-1 to the flowchart in Fig. 13-2, you ought to note a couple of important features. Note, for instance that the confidence level, variable KL, is carried as the third data element in the 5×5×3 memory array. This shows up most clearly in the second statement in line $2\emptyset5$, where KL=M(MI, MJ, 3). Line 245 in the S.$2\emptyset\emptyset$ subroutine then does the LOAD MEMORY operation, loading the new confidence value by the statement M(MI, MJ, 3)=KL.

As in earlier BETA programs, encoded values of I and J (expressed as numbers between 1 and 5) are saved in the first and second elements of the M array. See the first two statements in line 245. Load this S.$2\emptyset\emptyset$ BETA CONL (BETA CONL-1) subroutine from scratch:

```
200 REM ** BETA CONL-1 **
205 MI=I+3:MJ=J+3:KL=M(MI,MJ,3):IFKL=ØTHEN235
```

```
210 RI=M(MI,MJ,1):RJ=M(MI,MJ,2)
215 I=RI-3:J=RJ-3:GOSUB725
220 IFCO=ØTHEN24Ø
225 C=C+1:IFKL=ØTHEN235
230 KL=KL+1:IFKL=0THEN235ELSE215
235 GOSUB45Ø:GOSUB65Ø:KL=1:GOTO215
240 IFKL<4THENKL=KL+1ELSEKL=4
245 M(MI,M3,1)=I+3:M(MI,MJ,2)=J+3:M(MI,MJ,3)=KL:D=D+1:GOSUB
    1000. RETURN
```

Since the master program for BETA-I WITH CONFIDENCE is quite different from that of BETA-I BASIC, it should be loaded only after deleting the older master program. So do a DELETE for lines 1Ø through 9Ø before loading this program:

```
10 REM ** BETA CONL MASTER-1 **
15 DIMM(5,5,3):GOSUB5ØØ
20 CLS:FORN=1TO5·FORM-1TO5:FONO-1TO2:M(M,N,0)=3:NEXTO,M,N
25 C=C+1:MI=I+3:MJ=J+3:RI=M(MI,MJ,1):RJ=M(MI,MJ,2):IFNOT
   (RI=ØANDRJ=Ø)THEN35
30 GOSUB55:GOSUB6ØØ
35 X=X+I:Y=Y+J:GOSUB725
40 IFCO=ØTHEN5Ø
45 C=C+1:GOSUB45Ø:GOSUB2ØØ:GOSUB8ØØ
50 GOSUB45Ø:GOTO35
```

What You Can Expect From This Program

As mentioned earlier in this chapter, the confidence-level effect won't be noticeable through the creature's first few contacts with the border figure. In fact, the only time you will notice a confidence-level effect is when the creature has established a habit pattern of motion that becomes upset by an unexpected contact situation. Under this set of circumstances you will see the creature working at the same, non-working response several times in succession. Of course, the SCORE figure will drop by an unusual amount.

If you want to confirm that confidence levels are indeed being manipulated, enter this short testing program:

```
100 CLS:FORM=1TO5:FORN=1TO5:PRINTM",""N" "M(M,N,3)
105 INPUTS$:CLS:NEXTN,M:END
```

Run the program in the usual fashion, watching for the time the creature latches onto a habit pattern of motion. Then do a keyboard BREAK, followed by a GOSUB 1ØØ. This will call the little test routine, and you will see some figures on the screen. They will look something like this:

   2,3   3

The first two figures represent the address of the Beta memory as encoded versions of I and J respectively. In this particular example, 2,3 specifies motion-code address I=−1,J=∅.

The third figure is the confidence level associated with the motion-code address. Apparently the creature encountered a contact and solved the problems successfully, three different times. The solution, itself, isn't shown, but that could be included in the testing routine by adding a PRINT statement for M(M,N,1) and M(M,N,2).

You will be able to see the confidence level figure for all Beta memory locations by striking the ENTER key each time you want to see a new one.

Any confidence levels of zero indicate one of two things: Either the creature never encountered the situation or it failed to find a workable solution as often as it did find a good one.

If you have allowed the creature to execute its habit-pattern cycle several times, this little testing routine will show a few confidence levels of 4. Those places in Beta memory are the ones directly involved in the habit pattern. The remaining confidence levels, those between 1 and 3 inclusively, represent responses the creature executed in the process of establishing the habit pattern.

If you have any problems at all with the overall program, compare your listing with the composite version at the end of this chapter. And while you're at it, be on the lookout for statements in your program that aren't included in the composite listing. Delete any such program lines, thereby cleaning up the program memory for bigger things to come.

BETA-I CONFIDENCE WITH MANUAL DISTURB

The previous program does indeed represent the workings of a BETA-I creature with confidence-level features. It is difficult to appreciate the effects of adaptive trauma, however, because there are no provisions for actively upsetting the creature's habit pattern. With a few minor changes in the program, you can enter the picture and tease the little critter into breaking an old habit.

Figure 13-4 shows the scoring curves for a single adaptive sequence. The dashed curve shows how a BETA without the confidence-level feature responded to a new situation, and the solid curve shows how a BETA with confidence handled a situation that contradicted its past experience.

The BETA without the confidence feature recovered rather quickly since it readily rejected a remembered response that

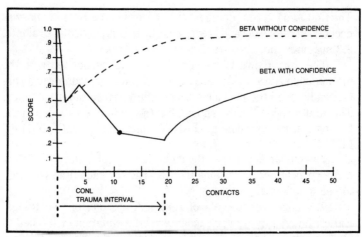

Fig. 13-4. Adaptive behavior of a BETA creature without the confidence feature and one possessing the confidence feature. Note the destruction of confidence through an adaptive trauma.

worked any number of times. The BETA creature with confidence, however, insisted upon trying the old habit response four times in succession, each try dropping the score a bit more. Once this BETA creature exhausted its confidence in the response (at about contact 20), it picked up a new one that worked. As a result, its scoring curve began showing signs of good learning taking place.

The curves in Fig. 13-4 demonstrate the vast differences in the adaptive behavior of BETAs running with and without the confidence-level feature. The data, however, represents a very small statistical sampling, and a critic of the idea could legitimately object to the nature of the curves on the basis that the sampling is so small.

Some first-hand experience with the program about to be described will convince you that the curves are quite representative of BETA behavior and adaptive trauma in particular. No one to date has run a compilation of a hundred or more such curves, and that's what is needed. So here's your chance to make a contribution to robotics.

The following program simply allows you to interrupt the normal sequence of events on the screen and inject a motion code of your choosing. For the sake of saving program memory space, you will find there are no goof-proofing features that automatically bypass invalid motion codes. The morale of the story is this: Make sure your motion codes are valid. Use Table 11-1 as a good listing of valid motion codes for this project.

Compare the flowchart in Fig. 13-5 with that in Fig. 13-1, and you will see the minor differences. The flowchart for this revised program includes a DISTURB conditional and an INPUT DISTURB operation. Whenever the creature is moving about freely on the screen, striking the *D* key interrupts the normal flow of events and calls the INPUT DISTURB subroutine, S.3ØØ. You will see from its listing that S.3ØØ simply allows you to enter new values of I and J (the motion code), set the creature figure near the center of the screen, set the scoring to zero, and resume activity after redrawing the field at DRAW FIELD.

Loading BETA-I CONFIDENCE WITH MANUAL DISTURB

Save the program described in the previous section on cassette tape, and then make the following revisions and additions to transform it into the MANUAL DISTURB version. First rewrite S.5ØØ HEADING this way:

```
500 REM ** BETA CONL DIST HEAD-1 **
505 CLS:PRINTTAB(25)"RODNEY BETA-I":PRINTTAB(2Ø)"CONL DIS-
TURB VERSION"
510 INPUT"DO ENTER TO START";S$:RETURN
```

Next, revise the master program to conform to this listing:

```
10 REM ** BETA CONL DIST MASTER-1 **
15 DIMM(5,5,3):GOSUB500
20 CLS:FORN=1TO5:FORM=1TO5:FORO=1TO2:M(M,N,0)=3:NEXTO,M,N
25 FORN=1TO5:FORM=1TO5:M(M,N,3)=Ø:NEXTM,N:GOSUB600
30 GOSUB55Ø
35 X=X+I:Y=Y+J:GOSUB725
40 IFCO=0THEN5Ø
45 C=C+1:GOSUB45Ø:GOSUB2ØØ:GOSUB8ØØ
50 GOSUB45Ø:IFINKEY$<>"D"THEN35
55 GOSUB3ØØ:GOTO3Ø
```

And finally, add this little S.3ØØ subroutine:

```
300 REM ** BETA DIST-1 **
305 CLS:INPUT"NEWI,J";I,J:CLS:X=45:Y=25:C=0:D=Ø:RETURN
```

The remainder of the subroutines are already in your program memory.

How to Use This Program

Run the program and give the BETA creature a chance to pick up its habit pattern of motion and then run that pattern for several cycles. Then, it is time to strike the *D* key and respond to the

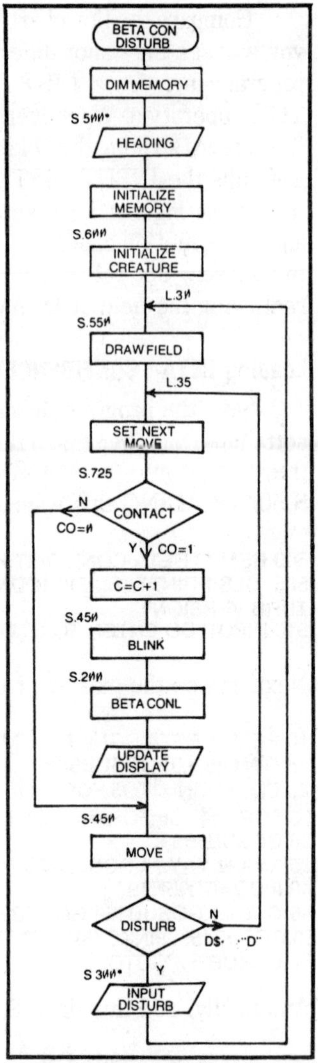

Fig. 13-5. Master program flowchart for BETA-I CONFIDENCE WITH MANUAL DISTURB.

request for new values of I and J. If you want to try the complete list of valid motion codes, use the listing in Table 11-1 as a guide.

Remember, this program cannot tolerate invalid motion codes. Entering values of I and J that are outside the bounds of -2 to 2 inclusively can run the creature outside its boundary, and when that happens the system goes into an error mode which can ultimately destroy the creatures hard-won memory contents. Just be sure to doublecheck your disturb motion code before entering it.

Each time you enter a new motion code, you have a chance to observe and record the creature's response to it. Sometimes it picks up the disturbance and accepts it without making any blunders at all. At other times, the new code upsets things a little bit, but if the creature finds the experience is one that doesn't have a very high confidence level, it recovers rapidly. There will always be a couple of disturbance motion codes that catch the creature in a habit pattern and throws his preconceived notions to the winds. It undergoes some violent trauma, as reflected by the scoring.

After you've had a chance to inject all 24 possible disturbance codes and write down the scoring history in each case, average the data to come up with a compiled response. The results ought to be something between the two kinds of curves shown in Fig. 13-4.

COMPILING ADAPTATION CURVES FOR BETA WITH CONFIDENCE

Inserting program steps that allow you to inject disturbance codes makes it possible to observe finely graded adaptive behavior on a first-hand basis. The procedure demands constant attention, and if you've done some of the work suggested in the previous section, you know that one complete set of tests requires 40 minutes to an hour of your time.

Such an experimental setup, one calling for your undivided attention, is not really suitable for extensive compiling projects. What is needed now is a set of long-term adaptation curves which will serve as a standard for the GAMMA work. In fact, the GAMMA projects don't mean very much if you do not have a set of BETA WITH CONFIDENCE curves at hand.

The idea is to devise an automated scheme for gathering adaptive data for BETA-I WITH CONFIDENCE. The general idea is to let a new creature go through its initial learning phase without any sort of outside interference. Once the habit pattern is established and has run at least four times, the next step is to slap it with the 24 disturbance phases listed in Table 11-1. Let the creature have 25 contacts to adapt to the new situation, recording the SCORE figure at the 5th and 25th contact in each case.

The situation is almost identical to that described for the BETA ADAPTATION COMPILE program in Chapter 11. The only difficulty is that smaller computer systems are running out of available memory space, and it is difficult to make this compiling program fully automatic.

So for the sake of experimenters using 4k systems, the compiling program in this section is semi-automatic. You start off the new creature and determine for yourself when it has established its habit pattern of motion. The general earmarks of this situation have been described several times in earlier BETA chapters.

The program is then set up so that it halts at contacts 5 and 25, giving you an opportunity to jot down the SCORE figure and, after

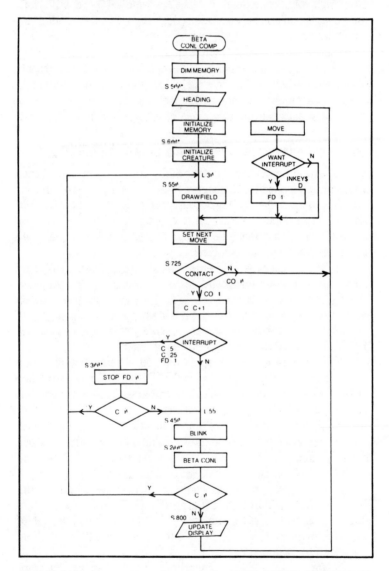

Fig. 13-6. Master program flowchart for BETA-I WITH CONFIDENCE COMPILE.

the 25th contact, enter a new disturbance motion code from the keyboard. The scheme does indeed call for your attention, but the fact that it automatically stops at the places requiring your full attention means it's possible to walk away from the experiment for a while without letting things get out of hand.

Flowcharts for BETA-I CONFIDENCE COMPILE

The master program for BETA-I CONFIDENCE COMPILE is shown in a flowchart form in Fig. 13-6. It is basically the same as the flowchart for the BETA-I CONFIDENCE WITH MANUAL DISTURB shown in Fig. 13-5, but there are a few differences that call for explanation.

The program starts out in the usual fashion, dimensioning the memory, doing a short HEADING routine, and initializing the memory and the creature. In fact, there are no significant differences through the main flow of events until the conditional, INTERRUPT, occurs.

INTERRUPT senses the time to stop all ongoing activity until you have a chance to gather the necessary data and tell the system what to do next. There are three interrupt conditions: C=5, C=25, and FD=1. Variable C, as usual, is defined as the number of contacts the creature has made during a given cycle. I've devised a system that interrupts the normal activity at the 5th and 25th contacts, so INTERRUPT must be sensitive to those two conditions.

The FD variable is a new one, though. It is a flag variable which causes an interrupt to occur anytime you specify, whether C happens to be equal to 5, 25, or anything else. The FD flag is actually set whenever conditional WANT INTERRUPT is satisfied. Note that satisfying WANT INTERRUPT brings up an FD=1 operation. As far as you, the operation, are concerned, WANT INTERRUPT is satisfied (and the interrupt flag, FD, is set) anytime you strike the D key while the creature is running freely. So, you can cause an interrupt by striking the D key, but normally you will want to let the interrupts occur automatically at C=5 or C=25.

At any rate, satisfying the INTERRUPT conditional brings up a new subroutine, S.300 STOP. During the STOP operation, the creature's activity is frozen on the screen and you can easily read the SCORE figure for the record.

The exact mechanism of the STOP subroutine will be described in a moment. For the time being, it is sufficient to say it is the operation that allows you to enter a disturbance code from the keyboard. If you specify it is time to do a new disturbance code, operations leave the STOP routine with C reset to zero. The C=∅ conditional is then satisfied and the system responds by redrawing the border figure and trying out the move you specified.

If, in the process of executing the STOP subroutine, you decide it is not time to inject a new motion code, variable C is *not* reset to zero; it retains the value it had upon entering the STOP subroutine. That being the case, C is not equal to zero, the C=∅ conditional is not satisfied, and the program returns to the normal BETA CONL loop.

There is a second C=∅ conditional on the flowchart, one directly following the S.2∅∅ BETA CONL subroutine. The reason

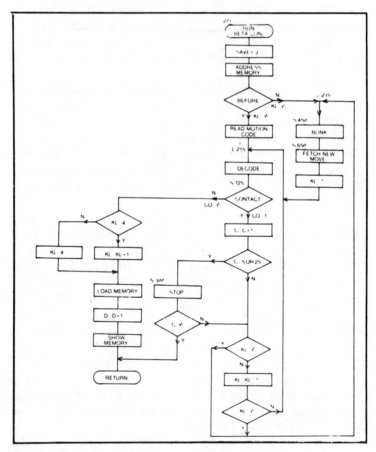

Fig. 13-7. Flowchart for subroutine 200 as revised for BETA-I WITH CONFIDENCE COMPILE.

for the second C=∅ conditional is that the new BETA CONL subroutine will, itself, contain and INTERRUPT conditional and STOP operation. So if the operations are stopped during BETA CONL, the second C=∅ conditional determines whether the figure should be restarted (a new disturbance code has been introduced) or resume its present activity (no new disturbance introduced during the BETA CONL subroutine).

The BETA CONL subroutine is flowcharted for you in Fig. 13-7. It is quite similar to the BETA CONL subroutine used in the two previous programs in this chapter. The main differences are the presence of two additional conditionals, C=5 OR C=25 and C=∅ and the subroutine STOP.

These additional operations amount to another interrupt routine, one that is almost identical to the interrupt routine in the master program of Fig. 13-6. The reason BETA CONL has to include this interrupt feature is that the contact-count variable often increments to 5 or 25 during the execution of this subroutine. If you were to omit the interrupt sequence here, the system might count through 5 or 25 contacts without sensing them in the master program.

The flowchart for the STOP subroutine is illustrated in Fig. 13-8. Whenever this subroutine is called by one of the interrupt conditions, either in the master program or in the BETA CONL subroutine, it first increments D, the NO. GOOD MOVES variable.

The fact that D is incremented upon doing the STOP sub-routine might sound like a bit of cheating in favor of the creature, but the process is necessary for showing scores of 1.000 when, indeed, they occur. Note that D is decremented by one at operation D=D−1 later in the flowchart, which means things are evened out in the long run anyway.

So D is incremented first and then the display is updated to reflect the current SCORE. The READY conditional is the one responsible for stopping all activity until you have a chance to record some data. When you've done that, simply striking the ENTER key satisfies the READY conditional and operations resume with yet another conditional operation—C *greater than or equal to* 25. If C is less than 25, variable D is restored to its original value (no cheating, you see), the border figure is redrawn on the screen to eliminate the question mark generated by the READY conditional, and control returns to the calling program.

Now, suppose C is equal to or greater than 25. In that case, the scheme calls the conditional, WANT DISTURB. In other words,

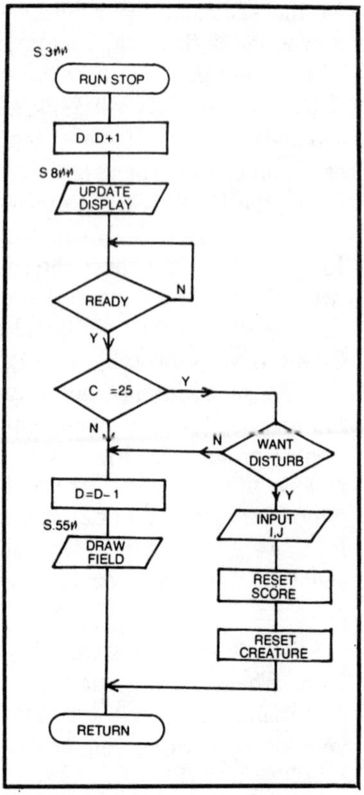

Fig. 13-8. Flowchart for subroutine
300 STOP.

"Do you want to enter a new motion code?" If not, the value of D is
adjusted to its original value by D=D−1 and things procede as
described before. But, if you do specify a disturb condition, the
system asks for the values of I and J you want to enter. This is
handled by INPUT I, J. Upon receiving those values from the
keyboard, the system resets the score to wipe out past scoring
history, sets the creature at a position near the middle of the
screen, and returns operations to the calling program, either the
main program or the S.2ØØ BETA CONL subroutine.

Loading BETA-I WITH CONFIDENCE COMPILE

This program can be loaded rather easily if the program from
the previous section is residing in program memory. Simply make
the following revisions:

```
300 REM ** STOP-1 **
305 D=D+1:GOSUB8ØØ:INPUTS$
```

```
310 IFO>=25THEN320
315 D=D-1:CLS:GOSUB550:RETURN
320 CLS:INPUT"DISTURB (Y OR N)";S$
325 IFS$="N"THEN315
330 INPUT"NEW I,J";I,J:C=0:D=0:X=45:Y=25:CLS:RETURN

200 REM ** BETA CONL-2 **
205 MI=I+3:MJ=J+3:KL=M(MI,MJ,3):IFKL=0THEN235
210 RI=M(MI,MJ,1):RJ=M(MI,MJ,2)
215 I=RI-3:J=RJ-3:GOSUB725
220 IFCO=0THEN240
225 C=C+1:IFC=50RC=25THEN250
227 IFKL=0THEN235
230 KL=KL-1:IFKL=0THEN235ELSE215
235 GOSUB450:GOSUB650:KL=1:GOTO215
240 IFKL<4THEN243
242 KL=4:GOTO245
243 KL=KL+1
245 M(MI,MJ,1)=I+3:M(MI,MJ,2)=J+3:M(MI,MJ,3)=KL:D=D+1:GOSUB1000:
    RETURN
250 GOSUB300:IFC=0RETURN
255 GOTO227

500 REM ** BETA CONL COMP HEAD-1 **
505 CLS:PRINTTAB(25)"RODNEY  BETA-I":PRINTTAB(23)"CONL  COMP
    VERSION"
510 INPUT"DO ENTER TO START":INPUTS$:RETURN

600 REM ** INITIAL COMP-4 **
605 GOSUB 650:X=RI+25:Y=25:GOSUB650:I=RI-3:J=RJ-3:C=0:D=0:
    RETURN
10 REM ** BETA CONL COMP-1 **
15 DIMM(5,5,3):GOSUB500
20 CLS:FORN=1TO5:FORM=1TO5:FORD=1TO2:M(M,N,O)=3:NEXTO,M,N
25 FORN=1TO5:FORM=1TO5:M(M,N,3)=0:NEXTM,N:GOSUB600
30 GOSUB550
35 X=X+I:Y=Y+J:GOSUB725
40 IFCO=0THEN65
45 C=C+1:IFC=5ORC=25ORFD=1THEN50ELSE55
50 FD=0:GOSUB300:IFC=0THEN30
55 GOSUB450:GOSUB200:IFC=0THEN30
60 GOSUB800
65 GOSUB450:IFINKEY$="D"FD=1
70 GOTO35
```

Running and Gathering Data From BETA CON COMP

This is a tedious process, or so it seems to use who have been
nurtured and spoiled by modern data-processing technology. Bear
in mind that the process comes to a halt when it is time for your to
do something, so you need not waste all your time watching the
program run. Why not study some more chapters in this book while
you keep an eye on the progress.

First run the program to launch a new creature. The program
will stop at the 5th contact and show the scoring, but chances are

quite remote that the critter has already established a habit pattern. So just strike the ENTER key and the program resumes. It will stop again at the 25th contact. Even then, there might not be a clear-cut habit pattern, so you will want things to continue as they are. Striking the ENTER key this time results in the question DISTURB (Y OR N)? You don't want to disturb the activity until the habit pattern has been established, so ENTER and *N* and you will see the activity picking up from where it left off.

Keep and eye on things until you see the creature has clearly established a habit pattern of motion. You know all the signs by now. Let it execute the pattern at least four times, then hit the *D* key. That's the only way you can get an interrupt to occur once the SCORE figure passes 25.

After striking the *D* key, the creature will continue crawling across the screen until it hits the border. At that moment the screen will clear and you'll get the question, DISTURB (Y OR N)? Now, you do indeed want to disturb the creature, so you should respond by entering Y. Doing so, you'll see the message NEW I,J? Enter your values of I and J that way, and I value, a comma, and a J value.

What value should you enter? Look at the worksheet in Fig. 13-9. The first suggested values are $-2,-2$. Enter these values and you'll see the border figure drawn afresh and the creature moving from the center of the border figure toward the upper left-hand corner. You won't see that the score has been reset until the creature makes its first contact with the border figure.

The creature adapts to its new motion code, the disturbance, in its own fashion until contact 5 occurs. When this occurs the program is halted and you should not the SCORE figure on the worksheet. Strike the ENTER key to resume.

Activity then procedes in the normal fashion until the creature makes its 25th contact of the cycle. Then it is time to again note the SCORE figure on the worksheet, respond to DISTURB (Y OR N) by entering Y, and respond to NEW I,J by entering the next set of motion-code values $--2,-1$ if this happens to be the second phase of the project.

Continue in this fashion until you have had a chance to enter all 24 possible combinations of disturbance motion codes and record the SCORE figure at the 5th and 25th in each case.

What next? Take a break, give your computer a keyboard BREAK and start all over again by entering RUN. Run as many of these cycles as you can remembering that it doesn't have to be done

PHASE	DISTURB		5	25	5	25	5	25	5	25	5	25	5	25	5	25
0																
1	−2	−2														
2	−2	−2														
3	−2	0														
4	−2	1														
5	−1	2														
6	−1	−2														
7	−1	−1														
8	−1	0														
9	−1	1														
10	−1	2														
11	0	−2														
12	0	−1														
13	0	1														
14	0	2														
15	1	−2														
16	1	−1														
17	1	0														
18	1	1														
19	1	2														
20	2	−2														
21	2	−1														
22	2	0														
23	2	1														
24	2	2														

Fig. 13-9. Suggested worksheet for gathering data from BETA-I WITH CONFIDENCE COMPILE.

in one day. The more critters you run, the more significant your results will be. When you've run all the creatures you want to run, average the results for each of the 24 phases at contacts 5 and 25. The averaged data can then be transferred to a graph, and there she is, a compiled adaptation curve for BETA-I creatures with confidence-level scoring.

Experimenters having access to computers with 16k or more of memory are certainly invited to compose a fully automatic version of this compiling experiment. At the time of this writing, no one in the world has done it yet.

COMPOSITE PROGRAM LISTINGS

BETA-I WITH CONFIDENCE Listing

```
10 REM ** BETA CONL MASTER-1 **

15 DIMM(5,5,3):GOSUB500

20 CLS:FORN=1TO5:FORM=1TO5:FORO=1TO2:M(M,N,O)=3:NEXTO,M,N

25 FORN=1TO5:FORM=1TO5:M(M,N,3)=0:NEXTM,N

30 GOSUB550:GOSUB600

35 X=X+I:Y=Y+J:GOSUB725

40 IFCO=0THEN50

45 C=C+1:GOSUB450:GOSUB200:GOSUB800

50 GOSUB450:GOTO35

200 REM ** BETA CONL-1 **

205 MI=I+3:MJ=J+3:KL=M(MI,MJ,3):IFKL=0THEN235

210 RI=M(MI,MJ,1):RJ=M(MI,MJ,2)

215 I=RI-3:J=RJ-3:GOSUB725

220 IFCO=0THEN240

225 C=C+1:IFKL=0THEN235

230 KL=KL-1:IFKL=0THEN235ELSE215

235 GOSUB450:GOSUB650:KL=1:GOTO215

240 IFKL<4THENKL=KL+1ELSEKL=4

245 M(MI,MJ,1)=I+3:M(MI,MJ,2)=J+3:M(MI,MJ,3)=KL:D=D+1:GO
    SUB1000:RETURN
450 REM ** BLINK-1 **

455 SET(X,Y):SET(X+1,Y):RESET(X,Y):RESET(X+1,Y)
```

264

```
460  RETURN

500  REM ** BETA CONL HEAD-1 **

505  CLS:PRINTTAB(25)"RODNEY BETA-1":PRINTTAB(25)"CONL
     VERSION"

510  INPUT"DO ENTER TO START";INPUTS$$:RETURN

550  REM ** FIELD-2 **

552  F1=15553:F2=15615:F3=16193:F4=16255

555  FORF=F1TOF2:POKEF,131:NEXT

560  FORF=F3TOF4:POKEF,176:NEXT

565  FORF=F1TOF4STEP64:POKEF,191:NEXT

570  FORF=F2TOF4STEP64:POKEF,191:NEXT

575  RETURN

600  REM ** INITIAL-2 **

610  X=10:Y=10:I=1:J=1

615  C=0:D=0

620  RETURN

650  REM ** FETCH NEW-1 **

660  RI=RND(5):RJ=RND(5)

665  IFRI=3ANDRJ=3THEN660

670  RETURN

725  REM ** CON SENSE-1 **

730  FORXP=2TO2+ABS(I):FORYP=1TO1+ABS(J)

735  IFPOINT(X+SGN(I)*XP,Y+SGN(J)*YP)=-1THEN750

740  NEXT:NEXT

745  CO=0:RETURN

750  CO=1:RETURN

800  REM ** UD SCORE-1 **

805  U$="#.###"

810  E=D/C

850  PRINT@15,"NO. CONTACTS"

852  PRINT@35,"NO. GOOD MOVES"

853  PRINT@55,"SCORE"

855  PRINT@87,C:PRINT@108,D
```

265

```
860 PRINT@119,USINGU$;E

865 RETURN

1000 REM ** SHOW MEM-1 **

1010 W=16327

1015 FORM=1TO5:FORN=1TO5:FORO=1TO2

1020 POKEW,127+M(M,N,O)

1025 W=W+1:NEXT:NEXT:NEXT

1030 RETURN
```

BETA-I CONFIDENCE WITH MANUAL DISTURB Listing

```
10 REM ** BETA CONL DIST MASTER-1 **

15 DIMM(5,5,3):GOSUB500

20 CLS:FORN=1TO5:FORM=1TO5:FORO=1TO2:M(M,N,O)=3:NEXTO,M,N

25 FORN=1TO5:FORM=1TO5:M(M,N,3)=0:NEXTM,N:GOSUB600

30 GOSUB550

35 X=X+I:Y=Y+J:GOSUB725

40 IFCO=0THEN50

45 C=C+1:GOSUB450:GOSUB200:GOSUB800

50 GOSUB450:IFINKEY$<>"D"THEN35

55 GOSUB300:GOTO30

200 REM ** BETA CONL-1 **

205 MI=I+3:MJ=J+3:KL=M(MI,MJ,3):IFKL=0THEN235

210 RI=M(MI,MJ,1):RJ=M(MI,MJ,2)

215 I=RI-3:J=RJ-3:GOSUB725

220 IFCO=0THEN240

225 C=C+1:IFKL=0THEN235

230 KL=KL-1:IFKL=0THEN235ELSE215

235 GOSUB450:GOSUB650:KL=1:GOTO215

240 IFKL<4THENKL=KL+1ELSEKL=4

245 M(MI,MJ,1)=I+3:M(MI,MJ,2)=J+3:M(MI,MJ,3)=KL:D=D+1:
    GOSUB1000:RETURN

300 REM ** BETA DIST-1 **

305 CLS:INPUT"NEW I,J";I,J:CLS:X=45:Y=25:C=0:D=0:RETURN
```

266

```
450 REM ** BLINK-1 **

455 SET(X,Y):SET(X+1,Y):RESET(X,Y):RESET(X+1,Y)

460 RETURN

500 REM ** BETA CONL DIST HEAD-1 **

505 CLS:PRINTTAB(25)"RODNEY BETA-I":PRINTTAB(20)"CONL
    DISTURB VERSION"
510 INPUT"DO ENTER TO START";S$:RETURN

550 REM ** FIELD-2 **

552 F1=15553:F2=15615:F3=16193:F4=16255

555 FORF=F1TOF2:POKEF,131:NEXT

560 FORF=F3TOF4:POKEF,176:NEXT

565 FORF=F1TOF4STEP64:POKEF,191:NEXT

570 FORF=F2TOF4STEP64:POKEF,191:NEXT

575 RETURN

600 REM ** INITIAL-2 **

610 X=10:Y=10:I=1:J=1

615 C=0:D=0

620 RETURN

650 REM ** FETCH NEW-1 **

660 RI=RND(5):RJ=RND(5)

665 IFRI=3ANDRJ=3THEN660

670 RETURN

725 REM ** CON SENSE-1 **

730 FORXP=2TO2+ABS(I):FORYP=1TO1+ABS(J)

735 IFPOINT(X+SGN(I)*XP,Y+SGN(J)*YP)=-1THEN750

740 NEXT:NEXT

745 CO=0:RETURN

750 CO=1:RETURN

800 REM ** UD SCORE-1 **

805 U$="$.###"

810 E=D/C

850 PRINT@15,"NO. CONTACTS"

852 PRINT@35,"NO. GOOD MOVES"
```

267

```
853 PRINT@55,"SCORE"

855 PRINT@87,C:PRINT@108,D

860 PRINT@119,USINGU$;E

865 RETURN

1000 REM ** SHOW MEM-1 **

1010 W=16327

1015 FORM=1TO5:FORN=1TO5:FORO=1TO2

1020 POKEW,127+M(M,N,O)

1025 W=W+1:NEXT:NEXT:NEXT

1030 RETURN
```

BETA CON COMP Program Listing

```
10 REM ** BETA CONL COMP-1 **

15 DIMM(5,5,3):GOSUB500

20 CLS:FORN=1TO5:FORM=1TO5:FORO=1TO2:M(M,N,O)=3:NEXTO,M,N

25 FORN=1TO5:FORM=1TO5:M(M,N,3)=0:NEXTM,N:GOSUB600

30 GOSUB550

35 X=X+I:Y=Y+J:GOSUB725

40 IFCO=0THEN65

45 C=C+1:IFC=5ORC=25ORFD=1THEN50ELSE55

50 FD=0:GOSUB300:IFC=0THEN30

55 GOSUB450:GOSUB200:IFC=0THEN30

60 GOSUB800

65 GOSUB450:IFINKEY$="D"FD=1

70 GOTO35

200 REM ** BETA CONL-1 **

205 MI=I+3:MJ=J+3:KL=M(MI,MJ,3):IFKL=0THEN235

210 RI=M(MI,MJ,1):RJ=M(MI,MJ,2)

215 I=RI-3:J=RJ-3:GOSUB725

220 IFCO=0THEN240

225 C=C+1:IFKL=0THEN235

230 KL=KL-1:GOTO215

235 GOSUB450:GOSUB650:KL=1:GOTO215
```

268

```
240  IFKL<4THENKL=KL+1ELSEKL=4
245  M(MI,MJ,1)=I+3:M(MI,MJ,2)=J+3:M(MI,MJ,3)=KL :D=D+1:
     GOSUB1000:RETURN
300  REM ** STOP-1 **
305  D=D+1:GOSUB800:INPUTS$
310  IFC>=25THEN320
315  D=D-1:CLS:GOSUB550:RETURN
320  CLS:INPUT"DISTURB (Y OR N)";S$
325  IFS$="N"THEN315
330  INPUT"NEW I,J";I,J:C=0:D=0:X=45:Y=25:CLS:RETURN
450  REM ** BLINK-1 **
455  SET(X,Y):SET(X+1,Y):RESET(X,Y):RESET(X+1,Y)
460  RETURN
500  REM ** BETA CONL COMP HEAD-1 **
505  CLS:PRINTTAB(25)"RODNEY BETA-1":PRINTTAB(23)"CONL COMP
     VERSION"
510  INPUT"DO ENTER TO START";INPUTS$:RETURN
550  REM ** FIELD-2 **
552  F1=15553:F2=15615:F3=16193:F4=16255
555  FORF=F1TOF2:POKEF,131:NEXT
560  FORF=F3TOF4:POKEF,176:NEXT
565  FORF=F1TOF4STEP64:POKEF,191:NEXT
570  FORF=F2TOF4STEP64:POKEF,191:NEXT
575  RETURN
600  REM ** INITIAL COMP-4 **
605  GOSUB650:X=RI+25:Y=RJ+25:GOSUB650:I=RI-3:J=RJ-3 :C=0:
     D=0:RETURN
650  REM ** FETCH NEW-1 **
660  RI=RND(5):RJ=RND(5)
665  IFRI=3ANDRJ=3THEN660
670  RETURN
725  REM ** CON SENSE-1 **
730  FORXP=2TO2+ABS(I):FORYP=1TO1+ABS(J)
735  IFPOINT(X+SGN(I)*XP,Y+SGN(J)*YP)=-1THEN750
740  NEXT:NEXT
745  CO=0:RETURN
750  CO=1:RETURN
800  REM ** UD SCORE-1 **
805  U$="$.###"
810  E=D/C
850  PRINT@15,"NO. CONTACTS"
852  PRINT@35,"NO. GOOD MOVES"
853  PRINT@55,"SCORE"
855  PRINT@87,C:PRINT@108,D
860  PRINT@119,USINGU$;E
865  RETURN
1000 REM ** SHOW MEM-1 **
1010 W=16327
1015 FORM=1TO5:FORN=1TO5:FORO=1TO2
1020 POKEW,127+M(M,N,O)
1025 W=W+1:NEXT:NEXT:NEXT
1030 RETURN
```

14 Introducing The Gamma-Class Creature

An Alpha-Class machine is one that shows only reflexive responses to changing environmental conditions. What's more, it lives just in the moment—it has no memory of past experience and it certainly cannot contemplate the future.

A Beta-Class machine can remember and use successful responses of the past to help solve problems of the present. Like its Alpha-Class ancestor, however, a Beta-Class creature has absolutely no recognition of future events.

A Gamma-Class machine represents the next logical step in this evolution of intelligent robot machines. Like its Beta-Class ancestor, a Gamma machine lives in the present and has recall of past events. The real hallmark of a Gamma creature, however, is its ability to anticipate events that have not yet occurred. It has the ability to enhance the quality of its present condition by drawing upon remembered experiences and generalizing them to future conditions.

The creation of a Gamma-Class creature follows a very logical pattern:

Alpha Class—knowledge of present conditions only

Beta Class—knowledge of past and present conditions

Gamma Class—knowledge of past, present, and future conditions.

I believe that any other evolutionary scheme would be needlessly complicated.

Now, recall that the BETA programs used ALPHA programs as starting points. Every BETA creature has a powerful ALPHA component in its modes of behavior. But, as a BETA creature learns its way around the world, the ALPHA component of its behavior becomes less obvious, giving way to remembered responses.

A GAMMA program is then an extension of a basic BETA program. It has some strong ALPHA characteristics at first, but given some experience with the world, these purely reflexive responses give way to BETA-like responses. However, given some BETA-like experience, which also fades into the background in favor of even higher-order GAMMA responses, problems never encountered before are solved using theories generated by past successes.

GENERALIZATION IS THE KEY TO GAMMA BEHAVIOR

A Gamma-Class machine is able to anticipate future problems by generalizing some of the relevant qualities of first-hand experiences. As an example, suppose you were taught on the blackboard in your grade school that 2 plus 2 equals 4. Each time the teacher wrote 2+2=? on that same blackboard you then knew the correct response is the numeral 4. You soon learned to solve the problem perfectly, using direct experiences of the past to solve a problem of the present.

But, what if the teacher then tests your learning by showing you the same problem written on a sheet of paper. Could you handle that? The problem is the same (2+2=?), but the situation for solving it is somewhat different—it's on a sheet of paper rather than the blackboard.

You certainly should be able to work the problem just as well. You have the ability to generalize first hand experiences capturing the essential elements of the problem so you can deal with it under a wide variety of different, but non-relevant, conditions in the future. The essential elements of this situation are that 2+2=4, and the solution is the same whether the problem is presented on the blackboard, a sheet of paper, or, for that matter, any other appropriate medium.

This sort of generalizing process sounds easy in the context of intelligent human beings. It isn't all that easy for a machine, however, Machines tend to be very literal in their interpretation of surrounding conditions. In order to recognize a problem once solved in the past, it must perceive every element of the environment as being exactly the same as before. Beta-Class machines work that way.

A Gamma-Class machine searches its memory for the relevant conditions of problems and then attempts to fit the relevant ideas into a framework of non-relevant surroundings that might occur in the future. It is thus prepared to deal with the situation under a variety of different conditions.

All of the ALPHA-I and BETA-I creatures described in this book are capable of recognizing and responding to 24 different environmental conditions. It might be argued that there are actually an infinite number of environmental conditions that can be solved using any of 24 possible responses. But no matter how you look at it, there are 24 kinds of relevant condition-response situations.

Now, suppose you have a BETA creature that has established a habit pattern of motion, and it is dealing with maybe 5 different condition-response situations quite effectively. In fact, the creature's responses are so effective that the confidence levels for those responses grows to a maximum of 4. The remaining responses have much lower confidence levels, however. Most of the remaining responses have confidence levels of zero.

What happens when you disturb the habit pattern? Well, if the disturbance calls for encountering a brand new environmental situation, the BETA creature sees a confidence level of zero and switches over to ALPHA-like behavior to come up with a workable response on the spot.

But now suppose this same BETA creature, the one that has established a habit pattern of motion made up of 5 different condition-response situations, has a built-in GAMMA mechanism. In this case, the system is sensitive to confidence levels of 4. The moment it finds a confidence level of 4, the GAMMA creature breaks down the situation into its most relevant components, searching its own memory for those same elements as they exist in bits and pieces at very low confidence levels (0 or 1, to be exact.) Upon discovering the relevant elements of a successful response residing in memory and carrying very low confidence levels, GAMMA inserts the successful elements into those places. What works quite well under one set of conditions ought to work just as well under similar conditions not yet encountered on a first-hand basis.

In a manner of speaking, GAMMA generates theories about what should work under untested conditions. The theories are well founded, being based upon elements of responses that have been quite successful in the past. A GAMMA creature performs this sort of routine on every response in memory that carries a confidence level of 4. Once the job is done, there are few condition-response places in memory left untouched.

Disturbing a GAMMA creature in such a way that it faces a situation for the first time does not automatically throw back into ALPHA-type behavior. A theory, or at least bits and pieces of a theory, exist in most places in a trained GAMMA memory. It has had a chance to presuppose what it should do when you disturb it.

272

So when you disturb its habit pattern of motion, it calls upon the theory and executes it.

If the theory works—if the creature's conjecture has been correct—GAMMA immediately solves the problem and increases its confidence level for that condition-response combination. What was first established by conjecture is now confirmed by first-hand experience. The creature had no need to resort to ALPHA-creature behavior.

Of course, there will be times when the creature's conjectures are not correct for one reason or another. A theory based upon first-hand, successful experiences of the past simple doesn't do the job. In that case, GAMMA is forced to resort to ALPHA behavior, picking up a random response that solves the problem. The new, randomly generated response soon becomes part of GAMMA's storehouse of relevant knowledge, and it is then used for modifying other conjectured responses.

A GAMMA creature is fully capable of changing its mind—the quality of its conjectured responses—as it gathers further first-hand experience. So GAMMA not only works to perfect its responses, but its conjectured responses as well. A BETA creature only works to perfect its overt, first-hand experiences, having no regard for possibilities of the future.

The bottom line is that a GAMMA creature is surperbly equipped to deal with new circumstances. It adapts quite readily to disturbances in its environment—it is prepared to do so. As a result, the adaptation curves for a GAMMA creature are smoother and rise more steadily than they do for a BETA creature.

FLOWCHARTS FOR GAMMA-I BASIC

Figure 14-1 is the flowchart for a GAMMA-I master program. It looks very much like a BETA-I WITH CONFIDENCE flowchart, and indeed the operations *are* much the same. Note the GAMMA CONL subroutine operation. This is a confidence-level subroutine which is almost identical to the one used in the previous chapter.

GAMMA CONL does one thing a BETA CONL subroutine does not, it sets a GAMMA FLAG variable (GF) whenever a confidence level somewhere in memory *first* reaches the maximum level of 4, and also whenever a confidence level in memory *first* drops from 4 to 3 because a previously good response no longer works. Look ahead to the GAMMA CONL flowchart in Fig. 14-2 if you want to see how that feature works. At any rate, GAMMA CONL in Fig. 14-1 sets confidence levels and a GAMMA FLAG whenever appropriate.

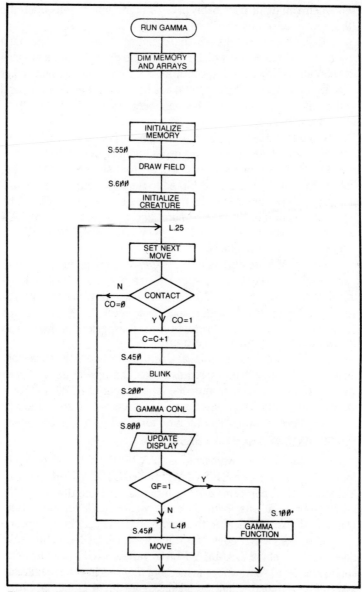

Fig. 14-1. Flowchart for GAMMA-I BASIC master program.

After doing the GAMMA CONL subroutine, the system updates the display on the screen by means of the usual S.800 UPDATE DISPLAY subroutine. Then the system checks the GAMMA FLAG.

If the GAMMA FLAG is not set (GF=∅), the whole program works as a BETA-I WITH CONFIDENCE (See Fig. 13-1). When the GAMMA FLAG is set, conditional GF=1 is satisfied and the program calls the GAMMA FUNCTION subroutine. This is the generalization subroutine that breaks down level-4 responses into their essential elements and assigns them to untested parts of the creature's memory.

If you did your homework and experiments with BETA-I WITH CONFIDENCE, this flowchart ought to seem rather simple,

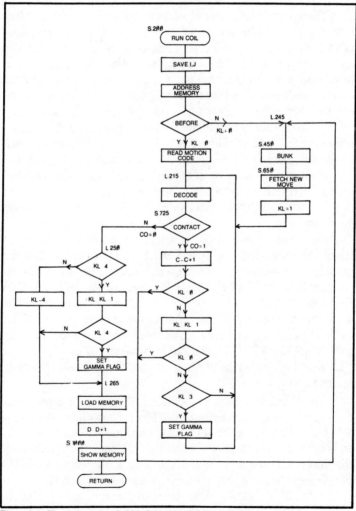

Fig. 14-2. Flowchart for S.200 GAMMA CONL. subroutine.

having only two really new operations—the GF=1 conditional and the GAMMA FUNCTION subroutine. You might also note that S.5ØØ HEADING has been omitted from the flowchart in Fig. 14-1. For smaller home computing systems, memory is getting scarce and some unnecessary flourishes must be eliminated.

The flowchart for GAMMA CONL is shown in Fig. 14-2. Compare it with the BETA version in Fig. 13-2, and you won't find very many differences. In fact, there are only four additional operations here: two SET GAMMA FLAG operations, a KL=4 conditional, and a KL=3 conditional.

There should be no need to explain the entire flowchart since that was done in Chapter 13. Of course, the additional operations call for some special consideration. Suppose the creature has found its way into the GAMMA CONL subroutine. According to the master program flowchart in Fig. 14-1, this happens every time the creature makes contact with the border figure. Further suppose the creature has indeed encountered this particular situation before (BEFORE conditional is satisfied), and the remembered response works (CONTACT conditional is *not* satisfied). This set of situations lead down to the KL *less than* 4 conditional in Fig. 14-2.

Is the confidence level of the remembered response less than 4? If so, KL=KL+1 increments the confidence level. It should do this because, you recall, the remembered response is working again this time.

Now, here is one of the new operations—the KL=4 conditional. The confidence level has just been incremented, and the question is whether or not the confidence level has been incremented to the highest possible level. If so, SET GAMMA FLAG sets variable GF equal to 1, and the system returns to the main program after doing several operations common to BETA CONL. If it turns out that the KL=4 conditional is *not* satisfied, the SET GAMMA FLAG operation is bypassed. Presumably the KL=KL+1 operation didn't carry the confidence level to the highest point yet.

The GAMMA FLAG is thus set to 1 whenever a particular remembered response reaches the 4 level. What's more this flag is set when the confidence level first reaches that level. If, after sensing no contact at the CONTACT conditional, the system finds that the confidence level is *not* less than 4, the entire GAMMA FLAG sequence is bypassed through the KL=4 operation.

Thus, the GAMMA FLAG is set to 1 when a certain remembered response shows a confidence level changing from 3 to 4. The flag routine is bypassed if that response is less than 4 or already-equal to 4 when the whole business starts.

There is a second GAMMA FLAG sequence in Fig. 14-2, however. This one is responsible for controlling the program when a confidence level decrements from 4 to 3. To see how this works, suppose the creature enters the GAMMA CONL subroutine with a situation calling for a response that has been successfully used before (BEFORE conditional is satisfied.) For some reason, however, that remembered response doesn't work, and the CONTACT conditional is satisfied. As a result, C=C+1 increments the NO. CONTACTS figure, and then the system wants to know whether or not the confidence level (KL) is already at zero. If not zero the confidence level is decremented at KL=KL-1 and checked again by the second KL=$\emptyset$ conditional. Assuming for the sake of this discussion that KL is not yet equal to zero, the system executes the KL=3 condition.

If KL is indeed equal to 3, the implication is that a previously high-confidence response is no longer working. It should not be used as part of the GAMMA generalization process, so the GAMMA FLAG is set to 1 by another SET GAMMA FLAG operation. Thus, the GAMMA FLAG is set whenever a confidence level drops from 4 to 3. Work your way around through this particular loop, and you will find that any other sort of drop in confidence level does not affect the GAMMA FLAG.

Aside from keeping track of confidence levels, GAMMA CONL is also responsible for setting the GAMMA FLAG to 1 under two conditions: whenever a confidence level *first* makes a transition from 3 to 4, and whenever a confidence level *first* makes a transition from 4 to 3. Under any other conditions, the subroutine works just like a BETA CONL version.

Now, return to the flowchart in Fig. 14-1 and you will be able to appreciate the significance of the GAMMA FLAG feature. In the flowchart, the GAMMA CONL subroutine is called whenever the creature makes contact with the border figure. After that, the GAMMA FUNCTION subroutine, S.1$\emptyset\emptyset$, is called only if GAMMA CONL has set the GAMMA FLAG to a value of 1. Otherwise, the GAMMA FUNCTION of the system is ignored altogether.

In other words, the GAMMA FUNCTION subroutine is called only two conditions: whenever a confidence level *first* makes a transition from 4 to 3 or from 3 to 4. The overall significance is that a GAMMA creature forms new theories about dealing with future events (executes the GAMMA FUNCTION subroutine) only when there is a marked change in some high confidence levels, either on the positive side (3-to-4 transitions) or on the negative side (4-to-3 transitions.)

The GAMMA FUNCTION subroutine is flowcharted in Fig. 14-3. While studying this flowchart, bear in mind that the subroutine is called only when there is some change at the top-levels of confidence. The direction of change, from 3 to 4 or from 4 to 3 is not relevant. Either case calls for a "change of mind."

The first operation in Fig. 14-3 is to print a pound sign (#) at the left-hand end of the memory-map display. In addition, this GAMMA-I program displays the creature's memory as most of the BETA programs did. Now, you will find a pound sign printed beside the memory map as long as the GAMMA FUNCTION operation is at work. There is a very good reason for showing the pound sign during the rather long GAMMA FUNCTION interval. Whenever GAMMA is "thinking" through its memory and devising theories, all creature activity on the screen comes to a halt. Without signalling the GAMMA FUNCTION in some fashion, you would have good cause to wonder whether or not something went wrong with the program.

After signalling the start of the GAMMA FUNCTION by printing the pound sign, the system clears something called a CONL COUNTER. As you will see in a moment, the CONL COUNTER keeps track of the number of level-4 confidence in the memory.

The system then clears something called the PHASE AC-CUMULATOR. Again, this will be described in greater detail a bit later in this section. It is sufficient to say right now that this operation clears a short-term memory which is responsible for accumulating the bits and pieces of knowledge that might have some relevance to future environmental conditions.

The next operation, INITIALIZE ADDRESS, is the first one of vital operational importance. The creature, you see, is going to search its entire memory for situations that have been solved before and carry confidence levels of 4. The search is going to be a very systematic one in this case, looking at every piece of memory data from the first to the last. This searching operation has to be initialized at the start, the justification for an INITIALIZING AD-DRESS operation.

The next step is a small subroutine calles S. 184 SET PHASE. This step classifies the address at hand according to some of its most relevant elements. In this particular generalization scheme there are eight different relevant contact situations:

1—current I and J motion parameters are both negative
2—current I parameter is negative, but J is zero
3—current I parameter is negative, but J is positive
4—current I parameter is zero, but J is negative
5—current I parameter is zero, but J is positive

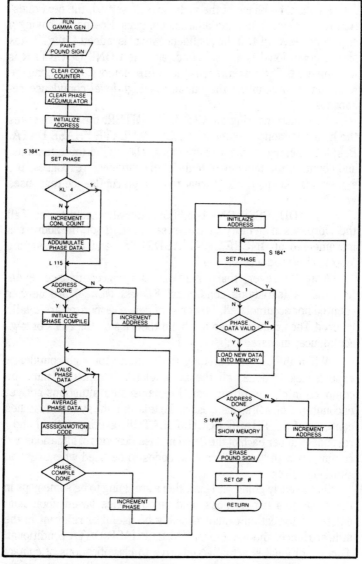

Fig. 14-3. Flowchart for S.100 GAMMA FUNCTION subroutine.

6—current I parameter is positive, but J is negative
7—current I parameter is positive, but J is zero
8—current I and J motion parameters are both positive.

That is how every conceivable contact situation is classified. The SET PHASE subroutine looks at the current address and classifies it according to one of those 8 different definitions.

So once the phase of the address is classified, the next question is whether or not the address contains a response having a confidence level of 4. If the confidence level is indeed 4, the KL *less than* 4 conditional is *not* satisfied, and the CONL COUNTER is incremented. The system scans the entire memory, classifying the situations and counting the number of high-level confidence responses.

After incrementing the CONL COUNTER, the system saves the high-confidence response at ACCUMULATE PHASE DATA. Besides scanning the entire memory, classifying the situations, and counting the number of high-level confidence responses, the system also keeps track of what those high-confidence responses are.

The ADDRESS DONE conditional merely checks to see if all the addresses in memory have been searched. If not, the address is incremented at INCREMENT ADDRESS , and the classifying loop is executed again.

Note, however, that conditions carrying confidence levels less than 4 (satisfying conditional KL *less than* 4) are neither counted nor accumulated. Lower-confidence responses are totally ignored. The creature does not build theories on the basis of low-confidence, untested experiences.

When the job of searching the memory for high-confidence experiences is done, all the accumulated information must be compiled into a set of suggested responses for situations not yet encountered on a first-hand basis. To this end, the system searches contents of the PHASE ACCUMULATOR, averaging the I and J responses under each of the 8 major classifications of situations and coming up with "typical" motion codes to be tried whenever the situation arises.

In the early going, at least, there are going to be some gaps in the creature's experiences, and any theories based upon non-existent information cannot possibly be useful or relevant in the future. Hence, the need for a VALID PHASE DATA conditional. This conditional is actually sensitive to classifications of contact situations that have not been experienced at all. As you might imagine, the number of such classifications tends to decrease with experience.

Upon finding some of this sort of invalid phase data, the system bypasses the two compiling steps, AVERAGE PHASE DATA and ASSIGN MOTION CODE and runs down to the point where the system looks for the end of the compiling operation. (See the PHASE COMPILE DONE conditional.) If it happens that the compiling operation has not been carried out for all eight possible classifications of contacts, the system increments the PHASE COUNTER at INCREMENT PHASE. (The PHASE COUNTER is actually the CONL COUNTER doing double duty.)

Reviewing what has taken place thus far, you can see that the GAMMA creature scanned its memory very systematically, classifying contact conditions, counting high-level responses, remembering those responses, and using them as a basis for compiling suggested motion codes where low confidence levels exist.

The final set of operations in Fig. 14-3 are responsible for loading those suggested motion codes into memory. To this end, the address is initialized and the contact conditions are again classified by subroutine S.184. This time, however, the system is looking for very low confidence levels, as opposed to high-confidence levels as in the first part of the routine.

Upon finding a confidence level of 1 or $\emptyset$ (by the conditional KL *greater than* 1 not being satisified), the system loads the appropriate components of the compiled motion code into the creature's memory. This operation is carried out by LOAD NEW DATA INTO MEMORY. However, the flowchart shows a conditional operation preceding this point—PHASE DATA VALID.

The I and J components of suggested motion codes are often entered separately—maybe an I component from one phase is combined with a J component from a different phase. There exists a distinct possibility that the two components make up a stop code (I=$\emptyset$ and J=$\emptyset$ at the same time). Since this particular motion code has been considered invalid for Level-I creatures elsewhere in this book, we simply follow the convention here.

So portions of the creature's memory that carry confidence levels of $\emptyset$ or 1 are loaded with suggested components of motion codes compiled in an earlier part of the subroutine. Responses carrying confidence levels of 2 or greater are not affected at all by the GAMMA FUNCTION subroutine.

Once the system has run through the entire addressing sequence, loading compiled or generalized motion codes into the appropriate places, it winds up operations by showing the new memory map, erasing the pound sign, resetting the GAMMA FLAG to zero and returning to the main program. All of this takes about 30 seconds.

LOADING GAMMA-I BASIC

It is probably pointless to dealve into the GAMMA process any further until you have had a chance to witness it in action for yourself. The program is best loaded with BETA-I WITH CONFIDENCE already residing in program memory. This being the case, you will find that the following subroutines can be used without any alterations:

```
S.450 BLINK or MOVE (BLINK-1)
S.550 DRAW FIELD (FIELD-2)
S.600 INITIALIZE CREATURE (INITIAL-2)
S.650 FETCH NEW MOVE (FETCH NEW-1)
S.725 CONTACT (CON SENSE-1)
S.100 SHOW MEMORY (SHOW MEM-1)
```

The GAMMA CONL subroutine, S.2ØØ GAMMA CONL-1, is quite different from the listing for its BETA counterpart. So delete lines 2ØØ through 245 and enter this GAMMA version:

```
200 REM ** GAMMA CONL-1 **
205 MI=I+3:MJ=J+3:KL=M(MI, MJ,3):IFKL= Ø THEN245
210 RI=M(MI,MJ,1):RJ=M(MI,MJ,2)
215 I=RI−3:J=RJ−3:GOSUB725
220 IFCO=Ø THEN25Ø
225 C=C+1:IFKL=ØTHEN24Ø
230 KL=KL−1:IFKL=Ø THEN245
235 IFKL=3THEN215
240 GF=1:GOTO215
245 GOSUB45Ø:GOSUB65Ø:KL=1:GOTO215
250 IFKL<4THEN255ELSE27Ø
255 KI=KL+1:IFKL=4THEN26Ø ELSE265
260 GF=1
265 M(MI,MJ,1)=RI:M(MI,MJ,2)=RJ:M(MI,MJ,3)=KL:D=D+1:
    GOSUB 1ØØØ:RETURN
270 KL=4:GOTO265
```

For the sake of saving memory space for more valuable operations, S.5ØØ HEADING (BETA CONL HEAD-1) should be deleted from your program memory. Do that by deleting lines 5ØØ through 51Ø.

Next, load GAMMA FUNCTION, S.1ØØ GAMMA GEN-1, from scratch:

```
100 REM ** GAMMA GEN-1 **
105 POKE16325, 35:FORK=1TO8:K(K)=∅:FORL=1TO2:G(K,L)=∅:NEXTL,
    K:FOR MI=1TO5:FORMJ=1TO5:GOSUB184:IFM(MI,MJ,3)<4THEN115
110 K(GP)=K(GP)=K (GP)+1:G(GP,1) =G(GP, 1)+M(MI, MJ, 1):G
    (GP,2)=G(GP,2)+M(MI, MJ, 2)
115 NEXT MJ, MI
120 FORGP=1TO8:FORGQ=1TO2
125 IFG(GP,GQ)=∅ORK(GP)=0THEN145
130 G(GP,GQ)=G(GP,GQ)/K(GP)
135 IFG(GP,GQ)<3THENG(GP,GQ)=2
140 IFG(GP,GQ)>3THEN(GP,GQ)=4
145 NEXTGQ,GP
150 GOTO16∅
155 G(GP,GQ)=∅:GOTO145
160 FORMI=1TO5:FORMJ=1TO5:GOSUB184
165 IFM(MI,MJ,3)>1∅RG(GP,1)=ORG(GP,2)=∅THEN175
170 M(MI,MJ,1)=G(GP,1):M(MI,MJ,2)=G(GP,2):M(MI,MJ,3)=1
175 NEXTMJ,MI
180 GOSUB1∅∅∅:POKE16325,32:GF=∅:RETURN
182 REM ** GAMMA PHASE-1 **
184 IFMI < 3ANDMJ < 3GP = 1
186 IFMI < 3ANDMJ =3GP=2
188 IFMI < 3ANDMJ > 3GP=3
190 IFMI = 3ANDMJ < 3GP=4
192 IFMI = 3ANDMJ > 3GP=5
194 IFMI > 3 ANDMJ < 3GP=6
196 IFMI > 3ANDMJ=3GP=7
198 IFMI > 3ANDMJ > 3GP=8
199 RETURN
```

Finally, replace the old master program with this one:

```
10 REM ** GAMMA-I MASTER-1 **
15 CLS:DIMM(5,5,3):DIMG(8,2):DIMK(8)
20 FORN=1TO5:FORM=1TO5:FORO=1TO3:M(M,N,O)=0:NEXTO,M,N:GO
   SUB55∅=GOSUB6∅∅
25 X=X+I:Y=Y+J:GOSUB 725
30 IFCO= ∅ THEN40
35 C=C+1:GOSUB45∅:GOSUB 2∅∅:GOSUB8∅∅:IFGF=1THEN45
40 GOSUB45∅:GOTO25
45 GOSUB1∅∅:GOTO25
```

WHAT TO EXPECT FROM GAMMA-I BASIC

When you first run this program, the border figure will appear on the screen almost immediately. There will be no HEADING message at all. There will be a brief delay, however, as the master program sets the dimensions of three memory arrays, the creature's main memory (array M(5,5,3)), the phase accumulator (array G(8,2)) and the CONL counter (array K(8)).

Once things get underway, the creature begins behaving in a typical BETA fashion. The moment it makes contact with the border figure, the scoring format appears on the screen, and the moment the creature picks a workable motion code, the memory map is drawn along the bottom of the screen. The creature bounces around in the normal way, eventually working its way into a habit pattern of motion—the whole thing appears to be working like a basic BETA-I operation.

However, after the GAMMA creature runs through its habit pattern several times you will note the first signs of truly Gamma-Class behavior. One of the responses in the habit pattern has picked up a confidence level of 4. Everything on the screen comes to a halt and a pound-sign character appears at the left-hand end of the memory map.

Watch that pound sign and memory map very carefully. About 15 seconds into the operation you will see portions of the memory map change. This will happen just before the GAMMA FUNCTION concludes and the creature goes about its business on the screen. The changes happen rather quickly and suddenly, and you don't want to miss out on seeing a computer reprogramming itself for the first time.

In most instances, the creature will resume its habit pattern, pausing for a GAMMA FUNCTION each time it meets one of the contact points. Once it has made it all the way through the habit pattern (all the involved responses have been set to a confidence level of 4 and the corresponding GAMMA FUNCTION has taken place), the thing continues running the pattern without any further GAMMA GEN operations. You will note, however, that few elements of the memory map are left untouched by the generalization processes.

The only way to see any further GAMMA activity is if the habit pattern is somehow disturbed. This particular program has no provisions for letting you step into the picture and upset the pattern, so you have to count on one of those situations where a certain response works for a number of times in succession, but suddenly fails to work one time.

The surest way to see some more GAMMA activity is to BREAK the program and start all over again. As you might suspect, the next chapter deals with ways to disturb the habit patterns and see the most significant effects of GAMMA activity—the ability to adapt to *foreseen* conditions not yet experienced on a first-hand basis.

SOME FURTHER NOTES ABOUT THE GAMMA FUNCTION SUBROUTINE

The GAMMA FUNCTION subroutine described in this chapter is just one of many possible approaches to generalizing the creature's experiences. Now that you have had a chance to study the flowchart and program listing, it is time to take a closer look at it. The GAMMA FUNCTION subroutine first breaks down the addresses of the memory into eight different categories. It systematically scans all the address locations, assigning a numerical value of 1 through 8 to variable GP.

The chart shown earlier in this chapter describes the nature of this address classification. GP is set to 1, for example, whenever the I and J values of the current motion code are both negative. But then, the same scheme gets GP equal to 2 whenever the I component of the current motion code is negative and the J component is equal to zero, and so on.

You should recall that memory is addressed by the motion code the creature is executing the moment a contact with the border figure occurs. Each address location then contains three items of data: the I and J components of a solution as well as the confidence level (KL) of that solution. Of course, that assumes a workable solution has been found before. If that isn't the case, all three items of memory data are zero.

Perhaps you should also be reminded that the I and J components of both the address and data are carried in an encoded form. Rather than being values between −2 and 2 inclusively (actual decoded I and J values), they are carried as integer values between 1 and 5.

At any rate, the first part of the GAMMA FUNCTION subroutine scans the entire family of memory address locations, assigning phase-related values of GP to the addresses *and* searching each address location for confidence levels of 4. Address locations containing confidence levels less than 4 are ignored. At this point the system is searching only for contact situations (addresses) that have been repeatedly solved (encoded I and J data) four times in succession (confidence level equals 4.)

Upon finding an address carrying confidence data equal to 4, the system does two things. First, it increments a confidence counter, a variable assigned to each of the eight different address phases. This variable is defind as K(GP). The second operation is to enter the encoded I and J response data into an accumulator. There are eight different accumulators, one for each of the eight different address phases and are defined as G(GP,GQ).

The phase accumulator is dimensioned at G(8,2) back in line 15 of the master program. The 8-item GP component carries the address phase classification number, while the 2-item GQ component picks up the values of I and J responses.

So whenever the system is scanning the memory addresses, searching for confidence levels of 4, it responds to finding that confidence level by incrementing a phase counter and saving the encoded value of the motion-code response in a phase accumulator. If a given address phase contains more than one 4-level confidence response, the corresponding sets of response data are summed together in the phase accumulator.

When this first address-scanning operation is completed, there are eight phase counters and eight phase accumulators containing response information of some sort. Those containing all zeroes are then ignored outright; there is no point in trying to generalize information that doesn't exist. However, the phase accumulators that do contain some response information are carrying sums of successful I and J responses. Furthermore, these sums of successful responses are organized according to the 8 possible address phases.

The phase accumulators containing valid information are then mathematically divided by their corresponding phase-counter values, thereby generating an average response code in each case. So what you have at this point is an average, or typical, motion code (a very high-confidence code) as possible solutions to various possible contact situations.

Now, the addresses, or possible contact situations, are classified by phases in the early going. By averaging the accumulated responses, they also take on a phase-related character. Suppose, for instance, that the accumulated I-value in a certain address phase accumulator has an average value of -1.5 (encoded value of $+1.5$) This is a clear indication that the successful responses built up for that particular address phase tend to have negative values of I, or encoded values less than 3. The situations represented by that particular address phase are thus generally solved by moving the creature from left to right. Of course, there might also be a corresponding J value to give the creature's motion a vertical component as well.

The averaging process is generally going to create some decimal values. These are corrected by assigning a particular integer value in lines 135 and 140 of the GAMMA FUNCTION subroutine. If an I or J value (encoded) is less than 3, the number 2 is substituted for it. An averaged motion component having a value less than 3 indicates a motion to the left for I components or a motion upward for J components. Substituting the value 2 simply sets a nice, in-the-middle, integer value that retains the proper phase without causing motions which are especially fast or slow.

The same sort of treatment applies to averaged response values greater than 3. The indication in this case is positive I or J values, and the system assigns a middle-of-the-road integer value of 4.

This marks the conclusion of the first part of the GAMMA FUNCTION subroutine featured in this book. What you have here is a set of typical responses to situations that have been encountered and solved successfully in the past. The next major step is to search the memory again, classifying the address phases just as before. This time, however, the system is looking for addresses carrying very low confidence levels, 0 or 1 to be specific.

Upon finding an address carrying a very low confidence level, the system looks at the typical responses to that sort of contact situation. If none are available from the first part of the job, the system goes on to the next address. The implication is that there is no confident information related to the situation at hand.

If, on the other hand, some compiled values of I or J are available for a low-confidence situation, those suggested values are loaded into the creature's memory and the confidence level is set at 1. The machine has successfully compiled a suggested motion code for one part of a particular address phase and inserted it as a possible motion code for a low-confidence situation within that same contact phase.

When the GAMMA FUNCTION subroutine is completely done, you have sets of high-confidence responses inserted into places of very low confidence within the same address phase. The system has taken the things it knows with great certainty and generalized it to similar circumstances where the creature has very low or, indeed, absolutely no certainity.

Note that the generalization process never raises the confidence level above 1. It can raise it from 0 to 1, but no higher than that. The purpose, here is to avoid the possibility of having the system build up very high confidence levels on the basis of conjecture alone. Confidence levels can be raised to 3 or 4 only by direct experience with the corresponding situations, and that takes place in the GAMMA CONL subroutine.

What's more, holding generalized confidence levels to a value of 1 allows the creature to adjust the generalized responses when new, high-confidence information becomes available. The generalization scheme, in other words, tends to optimize itself with additional experience in the environment.

But again, this is only one of many possible GAMMA FUNCTION schemes. The one described here is applicable to relatively small memories and a limited range of possible responses. You can be sure there are better schemes for larger systems. Alternate GAMMA GEN schemes are different, not only in their programming details, but also in their broader approach to the generalization problem.

One day soon, our lab here will be turning out a sophisticated Level-III GAMMA GEN scheme—one operating in conjuncton with two or more elements of adaptive intelligence. There will be one GAMMA system running the creature's overt activities, but that GAMMA system, itself, will be controlled by a separate BETA mechanism. In essence, the system will learn how to learn on its own.

COMPOSITE PROGRAM LISTING

```
10 REM ** GAMMA-i MASTER-1 **

15 CLS:DIMM(5,5,3):DIMG(8,2):DIMK(8)

20 FORN=1TO5:FORM=1TO5:FORO=1TO3:M(M,N,O)=0:NEXTO,
   M,N:GOSUB550:GOSUB600

25 X=X+I:Y=Y+J:GOSUB725

30 IFCO=0THEN40

35 C=C+1:GOSUB450:GOSUB200:GOSUB800:IFGF=1THEN45

40 GOSUB450:GOTO25

45 GOSUB100:GOTO25

100 REM ** GAMMA GEN-1 **

105 POKE16325,35:FORK=1TO8:K(K)=0:FORL=1TO2:G(K,L)
    =0:NEXTL,K:FORMI=1TO5:FORMJ
=1TO5:GOSUB184:IFM(MI,MJ,3)<4THEN115

110 K(GP)=K(GP)+1:G(GP,1)=G(GP,1)+M(MI,MJ,1):G(GP,
    2)=G(GP,2)+M(MI,MJ,2)

115 NEXTMJ,MI

120 FORGP=1TO8:FORGQ=1TO2

125 IFG(GP,GQ)=0ORK(GP)=0THEN145

130 G(GP,GQ)=G(GP,GQ)/K(GP)

135 IFG(GP,GQ)<3THENG(GP,GQ)=2

140 IFG(GP,GQ)>3THENG(GP,GQ)=4

145 NEXTGQ,GP

150 GOTO160
```

288

```
155 G(GP,GQ)=0:GOTO145

160 FORMI=1TO5:FORMJ=1TO5:GOSUB184

165 IFM(MI,MJ,3)>1ORG(GP,1)=0ORG(GP,2)=0THEN175

170 M(MI,MJ,1)=G(GP,1):M(MI,MJ,2)=G(GP,2):M(MI,MJ,3)=1

175 NEXTMJ,MI

180 GOSUB1000:POKE16325,32:GF=0:RETURN

182 REM ** GAMMA PHASE-1 **

184 IFMI<3ANDMJ<3GP=1

186 IFMI<3ANDMJ=3GP=2

188 IFMI<3ANDMJ>3GP=3

190 IFMI=3ANDMJ<3GP=4

192 IFMI=3ANDMJ>3GP=5

194 IFMI>3ANDMJ<3GP=6

196 IFMI>3ANDMJ=3GP=7

198 IFMI>3ANDMJ>3GP=8

199 RETURN

200 REM ** GAMMA CONL-1 **
205 MI=I+3:MJ=J+3:KL=M(MI,MJ,3):IFKL=0THEN245

210 RI=M(MI,MJ,1):RJ=M(MI,MJ,2)

215 I=RI-3:J=RJ-3:GOSUB725

220 IFCO=0THEN250

225 C=C+1:IFKL=0THEN240

230 KL=KL-1:IFKL=0THEN245

235 IFKL=3THEN215

240 GF=1:GOTO215

245 GOSUB450:GOSUB650:KL=1:GOTO215

250 IFKL<4THEN255ELSE270

255 KL=KL+1:IFKL=4THEN260ELSE265

260 GF=1

265 M(MI,MJ,1)=RI:M(MI,MJ,2)=RJ:M(MI,MJ,3)=KL:D=D+1:GOSU
    B1000:RETURN
270 KL=4:GOTO265

450 REM ** BLINK-1 **
```

289

```
455 SET(X,Y):SET(X+1,Y):RESET(X,Y):RESET(X+1,Y)
460 RETURN
550 REM ** FIELD-2 **
552 F1=15553:F2=15615:F3=16193:F4=16255
555 FORF=F1TOF2:POKEF,131:NEXT
560 FORF=F3TOF4:POKEF,176:NEXT
565 FORF=F1TOF4STEP64:POKEF,191:NEXT
570 FORF=F2TOF4STEP64:POKEF,191:NEXT
575 RETURN
600 REM ** INITIAL-2 **
610 X=10:Y=10:I=1:J-1
615 C=0:D=0
620 RETURN
650 REM ** FETCH NEW-1 **
660 RI=RND(5):RJ=RND(5)
665 IFRI=3ANDRJ=3THEN660
670 RETURN
725 REM ** CON SENSE-1 **
730 FORXP=2TO2+ABS(I):FORYP=1TO1+ABS(J)
735 IFPOINT(X+SGN(I)*XP,Y+SGN(J)*YP)=-1THEN750
740 NEXT:NEXT
745 CO=0:RETURN
750 CO=1:RETURN
800 REM ** UD SCORE-1 **
805 U$="#.###"
810 E=D/C                              1000 REM ** SHOW MEM-1 **
850 PRINT@15,"NO. CONTACTS"            1010 W=16327
852 PRINT@35,"NO. GOOD MOVES"          1015 FORM=1TO5:FORN=1TO5:
853 PRINT@55,"SCORE"                        FORO=1TO2
855 PRINT@87,C:PRINT@108,D             1020 POKEW,127+M(M,N,O)
860 PRINT@119,USINGU$;E                1025 W=W+1:NEXT:NEXT:NEXT
865 RETURN                             1030 RETURN
```

Testing Gamma-I's Adaptability

The program in Chapter 14 illustrates the workings of a basic GAMMA-I creature, but it does not afford any chance to study the superb adaptive qualities of a Gamma-Class machine. Hence, the purpose of this chapter.

The first program allows you to enter the GAMMA scheme at will, disturbing the creature's motion by entering any valid values of I and J from the keyboard. The second program runs the long-range, 24-phase compiling experiment, stopping for your attention at contacts 5 and 25 in each phase. Actually, these two programs parallel the two disturbance programs already described in Chapter 13.

The purpose of the first program in this chapter is to give you a chance to upset the creature's knowledge of the environment and observe the adaptive trauma in detail—if you can find any such thing. Most GAMMA creatures anticipate a good many disturbances, and you often have trouble finding one that isn't at least partly covered by an earlier generalization operation.

Aside from this primary purpose, the first program also lets you "train" a GAMMA creature. Indeed, it is defeating the whole philosophy of evolutionary adaptive machine intelligence to program any sort of overt behavior, but in a sense, the training procedures outlined for this GAMMA creature don't tell the creature exactly what it is supposed to do. Rather, it forces the creature to deal with one particular contact situation, and nothing else, until its confidence reaches the highest possible level. By selecting the training steps very carefully, you'll find that the creature can be taught to deal with all possible contingencies on the basis of a relatively few training steps.

The purpose of the second program in this chapter is to generate the primary adaptive learning curve for GAMMA-I creatures. Like its BETA CONL counterpart in Chapter 13, this is a long and tedious affair. However, the payoff is indisputable evidence that you are on the right track with evolutionary adaptive intelligence.

GAMMA-I WITH MANUAL DISTURB

This program, GAMMA-I WITH MANUAL DISTURB, is a variation of GAMMA-I BASIC presented in Chapter 14. To get things started, load GAMMA-I BASIC from your cassette tape and revise the master program to look like this one:

```
10 REM ** GAMMA-I MASTER-2 **
15 CLS:DIMM(5,5,3):DIMG(8,2):DIMK(8)
20 FORN=1TO5:FORM=1TO5:FORO=1TO3:M(M,N,O)=ØNEXTO,M,N:
22 GOSUB55Ø
25 X=X+I:Y=Y+J:GOSUB725
30 IFCO=ØTHEN4Ø
35 GOSUB8ØØ:IFGF=IFGF=1THEN45
40 GOSUB45Ø:IFINKEY$< >::D;;THEN25ELSE5Ø
45 GOSUB1ØØ:GOTO25
50 CLS:INPUT"NEW I,J";I,J:CLS:X=45:Y=25:C=Ø:D=Ø:GOTO22
```

That's all there is to the new program. Save it on cassette tape for later use. Then start having some fun with it.

Run the program and watch it work for awhile. Give the creature a chance to establish a habit pattern of motion and run through its cycles of GAMMA learning, learning operations signalled by a complete lack of creature activity on the screen and a pound sign appearing at the left-hand end of the memory map drawing on the screen. If you encounter any problems with the program, consult the composite listing at the end of this chapter.

After letting the creature learn its way around and do some generalizing of its confident know-how, strike the *D* key. This enables the disturbance mechanism. According to line 50 in the revised program, the screen will clear and the message NEW I, J will appear in the upper right-hand corner. Respond to the message with your selection of I and J values, typing them as an I value, followed by a comma, and the J value. You must enter valid values of I and J because the program has no provisions for goof-proofing your entry. Consult Fig. 13-9 for a complete listing of valid disturbance values.

As mentioned earlier, you might have some trouble finding a disturbance code that really upsets the GAMMA creature, providing you have given it ample opportunity to learn its way around initially. Finding it difficult to come up with a disturbance code that upsets the creature's scoring is not a bad thing, however. That's

292

what you want a GAMMA to do— take what it knows, generalize it to situations not yet encountered on a first-hand basis, then arrive at some theories concerning how those situations might be handled when they come about.

There is always a good chance that a GAMMA creature has some weaknesses in its storehouse of knowledge and experience. The problem is finding those weaknesses and using them to trip up the little monster. If you have trouble finding them, look at it this way—maybe you've created a creature that is smarter than you are. At least, it is elusive to the point where you're having trouble outsmarting it.

Use this program to gather some data and draw curves showing GAMMA-I working under stress. Find one of those disturbance codes that throw the creature into an adaptive trauma and gather as many figures as you can, showing the SCORE as a function of NO. CONTACTS.

TRAINING A GAMMA-I

It is possible to train a GAMMA-I creature to deal with future circumstances in a rational way. The idea is to start out with a fresh creature and before it has a chance to make its first border contact, interrupt its motion to enter a disturbance code from one of the eight fundamental GAMMA FUNCTION families. This kind of action on your part forces the creature to deal with that motion code, and not one picked on its own.

Disturb the creature's response, whatever it might be, before it has a chance to make another contact. Enter the same motion code and disturb it again before a contact occurs. Continue this training session until the creature does its GAMMA generalization operation, filling up the entire contact phase you have chosed with the one response it has been using.

Hitting the creature with the same motion code three times in succession, and interrupting the motion before a contact occurs, forces GAMMA generalization to occur in one major section of the memory. All other contact situations in that same phase take on the creature's selected response.

After the system completes that first generalization operation, and before the critter has a chance to make contact with the border figure, strike the *D* key again. This time, however, enter a motion code from a different contact phase. Continue this sequence until a new GAMMA operation occurs.

Here is a suggested list of motion codes for running this sort of training session:

$$I=-2, J=-2 \quad I=\emptyset, J=2$$
$$I=-2, J=\emptyset \quad I=2, J=-2$$
$$I=-2, J=2 \quad I=2, J=\emptyset$$
$$I=\emptyset \quad J=-2 \quad I=2, J=2$$

Included is one motion code from each of the contact phases in the GAMMA FUNCTION scheme. So if you force the creature to respond to each of these codes 3 times in succession, you will find the entire memory map filling up with information.

Then turn the little guy loose in his environment. Every response he makes will come from either a high-confidence, direct experience from the past or generalized information from such experiences. The creature might make some mistakes after you turn him loose; then again, he might not. It isn't a perfect creature, but after undergoing your extensive training routine, it is about as prepared for the world as it can be.

GAMMA-I WITH SEMI-AUTOMATIC COMPILE

By this time, the prospect of running a bunch of creatures through their paces and keeping track of their scores has become a routine matter. In this case, a GAMMA critter is allowed to learn his way around the world on the screen, and when it appears all the GAMMA functions have done their job, the creature is subjected to 24 sets of disturbance tests.

The chart back in Fig. 13-9 serves this purpose quite well, providing both the list of valid disturbance codes and some spaces for recording the SCORE figure at 5 and 25 contacts in each case.

There is little more to be added to the discussion about running BETA-I WITH CONFIDENCE COMPILE. This program runs the same way. This one includes the GAMMA FUNCTION of course, and the master program has to be rewritten to accommodate that feature.

There is one matter of very special importance to experimenters using systems having less than 16k of memory—YOU MUST ENTER A *CLEAR 16* BEFORE RUNNING THIS PROGRAM. Memory space is very tight on a 4K system and doing a CLEAR 16 is one tricky little way to pick up some more program memory space. You might have noticed in your Level II manual for the TRS-80 that the system automatically allocates 50 bytes for string memory space. It so happens that this program uses very little string space, so the space can be reduced to 16 by doing that CLEAR 16 command.

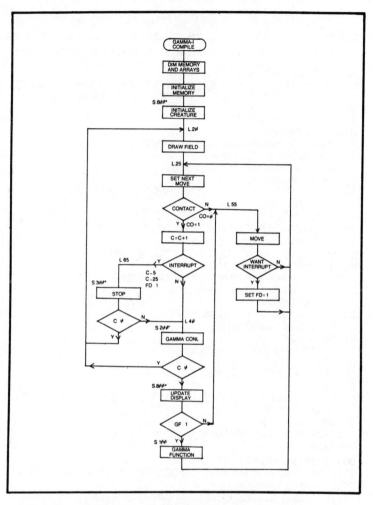

Fig. 15-1. Flowchart for GAMMA-I COMPILE master program.

In an effort to fit the program into a 4K computer format, some of the subroutines have been rewritten in shorter forms. Also, the BLINK subroutine must be eliminated and replaced by a different version inserted into the main program.

Figure 15-1 is the flowchart for the main program sequence. You should see that it is a rather straightforward combination of two flowcharts you have already studied in some detail: the master-program flowchart for GAMMA-I BASIC in Fig. 14-1 and the master-program flowchart for BETA-I CONFIDENCE COMPILE in Fig. 13-6.

Here are the listings for the portions of the program that are different from GAMMA-I BASIC in chapter 14. Load that program into your computer, then make these additions and modifications.

```
10 REM ** GAMMA-I MASTER-3 **
15 CLS:DIMM(5,5,3):DIMG(8,2):DIMK(8):FORN=1TO5:FORM=1 TO
   5:FORO=1T  Ø3:M(M,N,O)=Ø:NEXTO,M,N:GOSUB6ØØ
20 GOSUB55Ø
25 X=X+I:Y=Y+J:GOSUB725
30 IFCO=ØTHEN55
35 C=C+1:IFC=5ØORC=25ORFD=1THEN65
40 GOSUB2ØØ:IFC=ØTHEN2Ø
45 GOSUB8ØØ:IFGF=1THEN5ØELSE 55
50 GOSUB1ØØ:GOTO25
55 SET(X,Y):SET(X+1,Y):RESET(X,Y):RESET(X+1,Y):IFINKEY$=::D;;
   THEN6 ØELSE25
60 FD=1:GOTO25
65 FD=Ø:GOSUB3ØØ:IFC=ØTHEN2ØELSE4Ø
200 REM ** GAMMA CONL-2 **
205 MI=I+3:MJ=J+3:KL=M(MI,MJ,3):IFKL=ØTHEN245
210 RI=M(MI,MJ,1):RJ=M(MI,MJ,2)
215 I=RI-3:J=RJ-3:GOSUB725:IFCO=ØTHEN25Ø
220 C=C+1:IFC=5ØORC=25THEN275
225 IFKL=ØTHEN245
230 KL=KL-1:IFKL=ØTHEN245
235 IFKL=3THEN24ØELSE215
240 GF=1:GOTO215
245 GOSUB65Ø:KL=1:GOTO215
250 IFKL 4THEN255ELSE27Ø
255 KL=KL+1:IFKL=4THEN26ØELSE265
260 GF=1
265 M(MI, MJ, 1)=RI:M(MI, MJ, 2) = RJ:M (MI, MJ, 3) = KL: D=D+1: GOSUB
   ØØØ:RETURN
270 KL=:GOTO265
275 GOSUB3ØØ:IFC=ØRETURN
280 GOTO225
600 REM ** INITIAL-3A **
610 GOSUB65Ø:X=25+RI:Y=25+RJ:GOSUB65Ø:I=RI-3:J=3:RETURN
650 REM ** FETCH NEW-1A **
660 RI=RND(5):RJ=RND(5):IFRI=3ANDRJ=3THEN66Ø
670 RETURN
800 REM ** UD SCORE-1A **
850 PRINT@ 15'',NO. CONTACTS'':PRINT@35, "NO. GOOD MOVES"
   :PRINT@55,"SCORE":PRINT  87,C:PRINT  1Ø8,D:PRINT@119,USING
   "#.##";D/C:RETURN
1000 REM ** SHOW MEM-1A **
1010 W=16327:FORM=1 TO 5: FORN=1TO5: FORO=1 TO 2: POKEW, 127
    + M(M,N,O):W=W+1:NEXTO, N,M:RETURN
```

Now, compare your overall listing with the composite listing at the end of this chapter, deleting the lines that do not appear there. This deleting operation is absolutely essential if you are using a 4K system, and don't forget to enter a CLEAR 16 before attempting to run the program.

From the program for as many creatures as your time and patience allow. Average the results at contacts 5 and 25 for each of the 24 disturbance tests and then graph the results in the usual fashion.

You might want to keep the raw numbers, however. Some later time you will find it necessary to run the experiment a few more times and the new data can be compiled with the old, thereby creating a graph that has even better statistical significance.

COMPOSITE PROGRAM LISTINGS
GAMMA-I WITH MANUAL DISTURB Listing

```
10 REM ** GAMMA-i MASTER-2 **

15 CLS:DIMM(5,5,3):DIMG(8,2):DIMK(8)

20 FORN=1TO5:FORM=1TO5:FORO=1TO3:M(M,N,O)=0:NEXTO,M,N
   :GOSUB600
22 GOSUB550

25 X=X+I:Y=Y+J:GOSUB725

30 IFCO=0THEN40

35 C=C+1:GOSUB450:GOSUB200:GOSUB800:IFGF=1THEN45

40 GOSUB450:IFINKEY$<>"D"THEN25ELSE50

45 GOSUB100:GOTO25

50 CLS:INPUT"NEW I,J";I,J:CLS:X=45:Y=25:C=0:D=0:GOTO22

100 REM ** GAMMA GEN-1 **

105 POKE16325,35:FORK=1TO8:K(K)=0:FORL=1TO2:G(K,L)=0
    :NEXTL,K:FORMI=1TO5:FORMJ =1TO5:GOSUB184:IFM(MI,MJ,3)
    <4THEN115
110 K(GP)=K(GP)+1:G(GP,1)=G(GP,1)+M(MI,MJ,1)
    :G(GP,2)=G(GP,2)+M(MI,MJ,2)
115 NEXTMJ,MI

120 FORGP=1TO8:FORGQ=1TO2

125 IFG(GP,GQ)=0ORK(GP)=0THEN145

130 G(GP,GQ)=G(GP,GQ)/K(GP)

135 IFG(GP,GQ)<3THENG(GP,GQ)=2

140 IFG(GP,GQ)>3THENG(GP,GQ)=4

145 NEXTGQ,GP

150 GOTO160

155 G(GP,GQ)=0:GOTO145

160 FORMI=1TO5:FORMJ=1TO5:GOSUB184

165 IFM(MI,MJ,3)>1ORG(GP,1)=0ORG(GP,2)=0THEN175
```

```
170 M(MI,MJ,1)=G(GP,1):M(MI,MJ,2)=G(GP,2):M(MI,MJ,3)=1
175 NEXTMJ,MI
180 GOSUB1000:POKE16325,32:GF=0:RETURN
182 REM ** GAMMA PHASE-1 **
184 IFMI<3ANDMJ<3GP=1
186 IFMI<3ANDMJ=3GP=2
188 IFMI<3ANDMJ>3GP=3
190 IFMI=3ANDMJ<3GP=4
192 IFMI=3ANDMJ>3GP=5
194 IFMI>3ANDMJ<3GP=6
196 IFMI>3ANDMJ=3GP=7
198 IFMI>3ANDMJ>3GP=8
199 RETURN
200 REM ** GAMMA CONL-1 **
205 MI=I+3:MJ=J+3:KL=M(MI,MJ,3):IFKL=0THEN245
210 RI=M(MI,MJ,1):RJ=M(MI,MJ,2)
215 I=RI-3:J=RJ-3:GOSUB725
220 IFCO=0THEN250
225 C=C+1:IFKL=0THEN240
230 KL=KL-1:IFKL=0THEN245
235 IFKL=3THEN215
240 GF=1:GOTO215
245 GOSUB450:GOSUB650:KL=1:GOTO215
250 IFKL<4THEN255ELSE270
255 KL=KL+1:IFKL=4THEN260ELSE265
260 GF=1
265 M(MI,MJ,1)=RI:M(MI,MJ,2)=RJ:M(MI,MJ,3)=KL:D=D+1
    :GOSUB1000:RETURN
270 KL=4:GOTO265
450 REM ** BLINK-1 **
455 SET(X,Y):SET(X+1,Y):RESET(X,Y):RESET(X+1,Y)
460 RETURN
```

```
550 REM ** FIELD-2 **
552 F1=15553:F2=15615:F3=16193:F4=16255
555 FORF=F1TOF2:POKEF,131:NEXT
560 FORF=F3TOF4:POKEF,176:NEXT
565 FORF=F1TOF4STEP64:POKEF,191:NEXT
570 FORF=F2TOF4STEP64:POKEF,191:NEXT
575 RETURN
600 REM ** INITIAL-2 **
610 X=10:Y=10:I=1:J=1
615 C=0:D=0
620 RETURN
650 REM ** FETCH NEW-1 **
660 RI=RND(5):RJ=RND(5)
665 IFRI=3ANDRJ=3THEN660
670 RETURN
725 REM ** CON SENSE-1 **
730 FORXP=2TO2+ABS(I):FORYP=1TO1+ABS(J)
735 IFPOINT(X+SGN(I)*XP,Y+SGN(J)*YP)=-1THEN750
740 NEXT:NEXT
745 CO=0:RETURN
750 CO=1:RETURN
800 REM ** UD SCORE-1 **
805 U$="#.###"
810 E=D/C
850 PRINT@15,"NO. CONTACTS"
852 PRINT@35,"NO. GOOD MOVES"
853 PRINT@55,"SCORE"
855 PRINT@87,C:PRINT@108,D
860 PRINT@119,USINGU$;E
865 RETURN
1000 REM ** SHOW MEM-1 **
```

```
1010 W=16327

1015 FORM=1TO5:FORN=1TO5:FORO=1TO2

1020 POKEW,127+M(M,N,O)

1025 W=W+1:NEXT:NEXT:NEXT

1030 RETURN
```

GAMMA-I WITH SEMI-AUTO COMPILE Listing

```
10 REM ** GAMMA i MASTER-3 **

15 CLS:DIMM(5,5,3):DIMG(8,2):DIMK(8):FORN=1TO5:FORM=1
   TO5:FORO=1TO3:M(M,N,O)=0:NEXTO,M,N:GOSUB600

20 GOSUB550

25 X=X+I:Y=Y+J:GOSUB725

30 IFCO=0THEN55

35 C=C+1:IFC=50RC=250RFD=1THEN65

40 GOSUB200:IFC=0THEN20

45 GOSUB800:IFGF=1THEN50ELSE55

50 GOSUB100:GOTO25

55 SET(X,Y):SET(X+1,Y):RESET(X,Y):RESET(X+1,Y)
   :IFINKEY$="D"THEN60ELSE25

60 FD=1:GOTO25

65 FD=0:GOSUB300:IFC=0THEN20ELSE40

100 REM ** GAMMA GEN-1 **

105 POKE16325,35:FORK=1TO8:K(K)=0:FORL=1TO2:G(K,L)=0
    :NEXTL,K:FORMI=1TO5:FORMJ=1TO5:GOSUB184:IFM(MI,MJ,3)
    <4THEN115

110 K(GP)=K(GP)+1:G(GP,1)=G(GP,1)+M(MI,MJ,1):G(GP,2)=G
    (GP,2)+M(MI,MJ,2)

115 NEXTMJ,MI

120 FORGP=1TO8:FORGQ=1TO2

125 IFG(GP,GQ)=0ORK(GP)=0THEN145

130 G(GP,GQ)=G(GP,GQ)/K(GP)

135 IFG(GP,GQ)<3THENG(GP,GQ)=2

140 IFG(GP,GQ)>3THENG(GP,GQ)=4

145 NEXTGQ,GP

150 GOTO160
```

```
155 G(GP,GQ)=0:GOTO145

160 FORMI=1TO5:FORMJ=1TO5:GOSUB184

165 IFM(MI,MJ,3)>1ORG(GP,1)=0ORG(GP,2)=0THEN175

170 M(MI,MJ,1)=G(GP,1):M(MI,MJ,2)=G(GP,2):M(MI,MJ,3)=1

175 NEXTMJ,MI

180 GOSUB1000:POKE16325,32:GF=0:RETURN

182 REM ** GAMMA PHASE-1 **

184 IFMI<3ANDMJ<3GP=1

186 IFMI<3ANDMJ=3GP=2

188 IFMI<3ANDMJ>3GP=3

190 IFMI=3ANDMJ<3GP=4

192 IFMI=3ANDMJ>3GP=5

194 IFMI>3ANDMJ<3GP=6

196 IFMI>3ANDMJ=3GP=7

198 IFMI>3ANDMJ>3GP=8

199 RETURN

200 REM ** GAMMA CONL-2 **

205 MI=I+3:MJ=J+3:KL=M(MI,MJ,3):IFKL=0THEN245

210 RI=M(MI,MJ,1):RJ=M(MI,MJ,2)

215 I=RI-3:J=RJ-3:GOSUB725:IFCO=0THEN250

220 C=C+1:IFC=5ORC=25THEN275

225 IFKL=0THEN245

230 KL=KL-1:IFKL=0THEN245

235 IFKL=3THEN240ELSE215

240 GF=1:GOTO215

245 GOSUB650:KL=1:GOTO215

250 IFKL<4THEN255ELSE270

255 KL=KL+1:IFKL=4THEN260ELSE265

260 GF=1

265 M(MI,MJ,1)=RI:M(MI,MJ,2)=RJ:M(MI,MJ,3)=KL
    :D=D+1:GOSUB1000:RETURN
270 KL=4:GOTO265
```

301

```
275  GOSUB300:IFC=0RETURN

280  GOTO225

300  REM ** STOP-1 **

305  D=D+1:GOSUB800:INPUTS$

310  IFC>=25THEN320

315  D=D 1:CLS:GOSUB550:RETURN

320  CLS:INPUT"DISTURB (Y OR N)";S$

325  IFS$="N"THEN315

330  INPUT"NEW I,J";I,J:C=0:D=0:X=45:Y=25:CLS:RETURN

550  REM ** FIELD-2 **

552  F1=15553:F2=15615:F3=16193:F4=16255

555  FORF=F1TOF2:POKEF,131:NEXT

560  FORF=F3TOF4:POKEF,176:NEXT

565  FORF=F1TOF4STEP64:POKEF,191:NEXT

570  FORF=F2TOF4STEP64:POKEF,191:NEXT

575  RETURN

600  REM ** INITIAL-3A **

610  GOSUB650:X=25+RI:Y=25+RJ:GOSUB650:I=RI-3:J=RJ-3:RETURN

650  REM ** FETCH NEW-1A **

660  RI=RND(5):RJ=RND(5):IFRI=3ANDRJ=3THEN660

670  RETURN

725  REM ** CON SENSE-1 **

730  FORXP=2TO2+ABS(I):FORYP=1TO1+ABS(J)

735  IFPOINT(X+SGN(I)*XP,Y+SGN(J)*YP)=-1THEN750

740  NEXT:NEXT

745  CO=0:RETURN

750  CO=1:RETURN

800  REM ** UD SCORE-1A **
850  PRINT@15,"NO. CONTACTS":PRINT@35,"NO. GOOD MOVES"
     :PRINT@55,"SCORE":PRINT@ 87,C:PRINT@108,D:PRINT@119,
     USING"#.##";D/C:RETURN.
1000 REM ** SHOW MEM-1A **
1010 W=16327:FORM=1TO5:FORN=1TO5:FORO=1TO2:POKEW,
     127+M(M,N,O):W=W+1:NEXTO,N,M:RETURN
```

302

A Crash Course In EAMI Programming

By the time you get to this part of the work, you should have a pretty good idea about how EAMI works and how the programs themselves demonstrate EAMI and offer chances to discover new things about it.

The main purpose of the programs in this chapter is to pull together the complete EAMI system as it has been presented so far in this book. The programs are much tighter, run faster, and work more efficiently than any of the others. There are no special frills, however, so you will have to understand everything that's gone before in order to understand what is happening here.

The advantage of using the programming scheme in this chapter is that it lets you get a complete set of EAMI demonstration programs compacted into a relatively small section of cassette tape. You will then be able to load the programs from the tape in rather rapid succession, perhaps getting a clearer overall view of the EAMI process than is possible with the more elaborate versions.

The programs appear here with virtually no theory of operation. By this time, you have had enough experience with the basic EAMI concepts and programming procedures to run through the whole thing without making a lot of explanatory detours.

ALPHA-IG

Figure 16-1 shows the flowchart for a highly compacted version of ALPHA-I. If you get a little lost in your own analysis of the theory of operation, check the work against a similar sort of approach for ALPHA-II CONVERSION DEMO in Chapter 8.

This program must be entered all the way and from scratch. While the subroutine numbers will appear quite familiar, you will note that the statements are written somewhat differently. The use of a lot of POKE graphics makes it important to save the program on cassette tape before trying it on the machine.

```
10 REM ** ALPHA-IG MASTER **
15 CLS:PRINT@ Ø,"*ALPHA-I*":GOSUB55Ø:GOSUB6ØØ
20 GOSUB725:IFCO=32THEN35
25 C=C+1:GOSUB65Ø:I=RI−2:J=RJ−:GOSUB725:IFCO<>32 THEN 25
30 D=D+1:GOSUB8ØØ
35 GOSUB4ØØ:IFINKEY$="D"THEN1ØELSE2Ø

400 REM ** MOVE II-1 **
405 POKENP, 32:NP=NP+I+64*J:POKENP,14Ø:RETURN

550 REM ** FIELD II-1 **
555 FORF=15553TO15615:POKEF,131:NEXT
560 FORF=16193TO16255:POKEF,176:NEXT
565 FORF=15553TO1655STEP64:POKEF,191:NEXT
570 FORF=15615TO16255STEP64:POKEF,191:NEXT
575 RETURN

600 REM ** INITIAL IIG-1 **
605 NP=15968:GOSUB65Ø:I=RI−2:J−RJ−2:C−Ø:D=Ø:RETURN

650 REM ** FETCH NEW IIG-1 **
655 RI=RND(5)−1:RJ=RND(5)−1:IFRI=2ANDRJ=2THEN655
660 RETURN

725 REM ** CON SENSE II-1 **
730 FORSI=ABS(I)TO1STEP−1:FORSJ=ABS(J)TO1STEP−1
735 CO=PEEK (NP+SI*SGN(I)+64*SJ*SGN(J))
740 IFCO<>32RETURN
745 NEXTSJ,SI:RETURN

800 REM ** UD SCORE IIG-1 **
810 PRINT @ 13, @ "NO. CONTACTS":PRINT @ 3Ø, "NO. GOOD MOVE-
    S":PRINT@ 5Ø,"
815 PRINT $79,C:PRINT@ 99, D:PRINT @114, USING"###";D/C:PRINT
    125,CH R$(C1)
820 RETURN
```

Once you get the program running, you'll find you can disturb
the creature's activity and reset the scoring figures by simply
striking the *D* key.

ALPHA-IIG

The flowchart for an ALPHA-II Program appears in Fig. 16-2.
It is a no-nonsense version of the ALPHA-II CONVERSION
DEMO program in Chapter 8. Like the ALPHAIG program described in the previous section of this chapter, the creature can be
disturbed at any time by striking *D* key. The obstacles, incidentally, are the same ones used in earlier Level-II programs.

The program can be loaded into your computer by building it
around the ALPHA-IIG program. Simple revise the master program and add the figure-drawing subroutine as shown here:

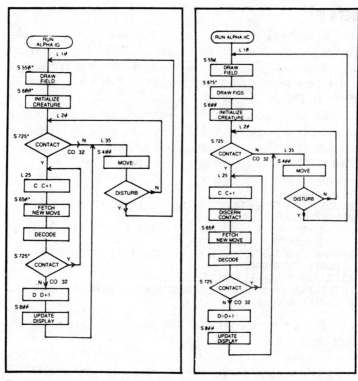

Fig. 16-1. Flowchart for ALPHA-IG. Fig. 16-2. Flowchart for ALPHA-IIG.

```
10 REM ** ALPHA-IIG MASTER **
15 CLS:PRINT@0, "*ALPHA-II*":PRINT@60, "FIG."15:GOSUB550:GOSUB
   675:GOSUB600
20 GOSUB725:IFCO=32THEN35
25 C1=CO:C=C+1:GOSUB650:I=RI−2:J=2:GOSUB725:IFCO
   <>°32THEN25
30 D=D+1:GOSUB 00
35 GSOBU400:IFINKEY$="D"THEN10ELSE20

675 REM **FIGS II-1**
677 P=15755
680 FORF=0TO1:FORN=P+64*F+8:POKEN,35:NEXTN,F
682 FORF=0TO1:FORN=P+32+64*FTOP+40+64*F:POKEN,48:NEXTN,F
684 FORF=0TO1:FORN=P =192 + 64* FTOP + 200 + 64* F: POKEN,42:
   NEXTN,F
686 FORF=0TO1:FORN=P+224+64*FTOP+232+64*POKEN,37:NEXTN,F
690 RETURN
```

Save it on cassette tape, and you're ready to go. If you run
into any difficulties, compare your program listing with the
composite version at the end of this chapter.

BETA-IG

You can now extend this logical sequence of program evolution by preparing the compacted version of BETA-I. The flowchart is in Fig. 16-3. Though this is a BETA-I creature program, it follows the general procedures outlined for BETA-II CONVERSION DEMO in Chapter 12.

The simplest way to get this new program into your machine is by building it around the ALPHA-IG version presented in the first section of this chapter. So load ALPHA-IG from cassette tape and then modify the master program as follows:

```
10 REM ** BETA-IG MASTER **
15 DIMM(4,4,2)
20 CLS:PRINT Ø,"*BETA-I*":GOSUB55Ø:GOSUB6ØØ
25 GOSUB725:IFCO=32THEN45
30 C=C+1: MI=I+2: MJ=J+2:RI=M(MI,MJ,1):RJ=M(MI,
   MJ,2):KL=M(MI,MJ,Ø):IFKL=ØTHEN 5Ø
35 I=RI−2:J=RJ−2:GOSUB725:IFCO<>32THEN55
40 D=D+1:GOSUB8ØØ
45 GOSUB4ØØ:IFINKEY$="D"THEN2ØELSE25
50 GOSUB65Ø:M(MI,MJ,1)=RI:M(MI,MJ,2)=RJ:M(MI,MJ,Ø)=1:GOTO35
55 C=C+1:GOTO5Ø
```

Again, save the program on cassette tape before actually running it.

You might have noticed from the flowchart (and you'll certainly notice when you run the program) that there is no SHOW MEMORY operation. This particular operation, used with virtually all the BETA programs described earlier in this book, burns up a lot of program execution time and thus slows down the action of the creature. Omitting this operation here lets the creature work faster and develop its habit pattern of motion a lot sooner.

BETA-IIG

This method of using highly compacted programs shows its real advantage in the case of this BETA-II program. The BETA-II programs presented in earlier chapters were capable of sensing the qualities of only two different obstacles and the border figure. The program space gained by using compacted programs allows the creature to sense and respond differently to all the figures on the screen.

This BETA-II program can be loaded on top of the ALPHA-IIG program outlined earlier in this chapter. Doing it this way, you only have to modify the master program and add the DISCERN CONTACT sensing subroutine, S.9ØØ. They are:

```
10 REM ** BETA-IIG MASTER **
15 DIMM(4,4,4,2)
```

20 CLS:PRINT @Ø,''*BETA-II*'':PRINT @6Ø,''FIG.'':GOSUB55Ø:GOSUB675:
   GOSUB6ØØ
25 GOSUB725:IFCO=32THEN45
3Ø C=C+1:C1=CO:GOSUB9ØØ:MI=I+2:MJ=J+2:RI=M(MI,MJ,B,1):RJ=M
   (MI,MJ ,B,2):IFM(MI,MJ,B,Ø)=ØTHEN5Ø
35 I=RI−2:J=RJ−2:GOSUB725:IFCO 32THEN55
4Ø D=D+1:GOSUB8ØØ
45 GOSUB4ØØ:IFINKEY$=''D''THEN2ØELSE25
5Ø GOSUB65Ø:M(MI,MJ,B,1)=RI:M(MI,MJ,B,2)=RJ:M(MI,MJ,B,Ø)=1:
   GOTO35
55 C=C+1:GOTO5Ø

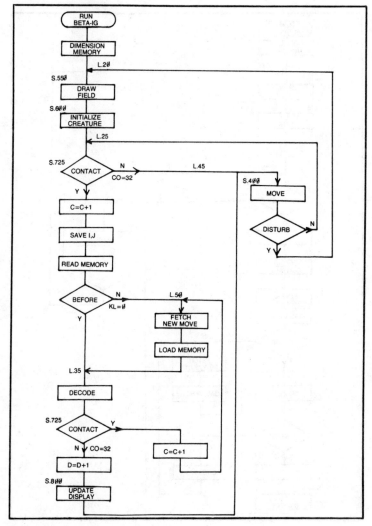

Fig. 16-3. Flowchart for BETA-IG.

307

```
900 REM ** CON EVAL IIG-1 **
905 IFCO=131ORCO=176ORCO=191THEN930
910 IFCO=35THEN935
915 ICFO=48THEN940
920 IFCO=42THEN945
925 IFCO=37THEN950
930 B=0:RETURN
935 B=1:RETURN
940 B=2:RETURN
945 B=3:RETURN
950 B=4:RETURN
```

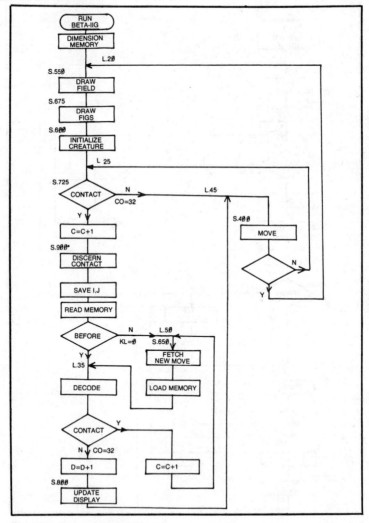

Fig. 16-4. Flowchart for BETA-IIG.

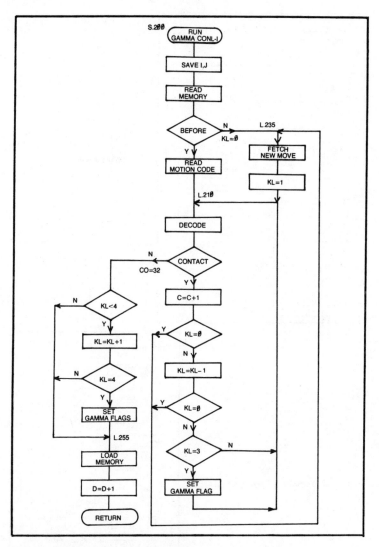

Fig. 16-5. Flowchart for GAMMA CONL.

Save the program on cassette tape and run it. You'll find it working beautifully. (Murphy's Law actually forbids such a statement, however, so you might want to check your complete listing against the composite version at the end of the chapter).

Let the critter establish its unique habit pattern of motion, then upset it by striking the D key. Chances are good that it will eventually find its way back to the same pattern and position on the

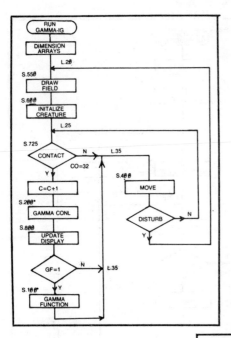

Fig. 16-6. Flowchart for GAMMA-IG.

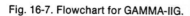

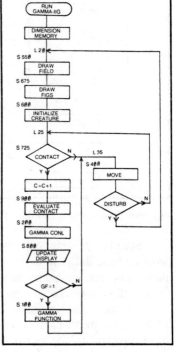

Fig. 16-7. Flowchart for GAMMA-IIG.

310

screen, clearly demonstrating how a sensitivity to the differences in quality between obstacles serve as cues to where the creature happens to be moving on the screen.

GAMMA-IG

Building a souped-up version of the GAMMA creature program calls for deleting some operations in the GAMMA CONL (confidence-level) subroutine. See the new GAMMA CONL routine flowcharted in Fig. 16-5.

The program statements for the GAMMA FUNCTION subroutine are also quite a bit different from those described in Chapter 14, but the flowchart operations aren't significantly different. The flowchart in Fig. 16-6 represents the overall activity of this condensed GAMMA-I program.

This program can be entered into the computer using the BETA-IG programming as a starting point. Simply rework the master program and enter the two new subroutines as shown here:

```
10 REM ** GAMMA-IG MASTER **
15 DIMM(4,4,2):DIMG(7,1):DIMK(7)
20 CLS:PRINT @Ø,"*GAMMA-I*":GOSUB55Ø:GOSUB6ØØ
25 GOSUB725:IFCO=32THEN35
30 C=C+1:GOSUB2ØØ:GOSUB8ØØ:IFGF=1THEN4Ø
35 GOSUB4ØØ:IFINKEY$="D"THEN2ØELSE25
40 GOSUB1ØØ:GOTO35
100 REM ** GAMMA-GEN G1 **
105 POKE16325,35:FORK= ØTO7:K(K)=Ø:FORL=ØTO1:G(K,L)=Ø:NEXTL,
    K: FOR MI= Ø TO4:FORMJ= Ø TO4:GOSUB175:IFM(MI,MJ,Ø)
    4THEN115
115 NEXTMJ,MI
120 FORGP=ØTO7:FORGQ=ØTO1
125 IFK(GP)=ØTHEN145
130 G(GP,GQ)=G(GP,GQ)/K(GP)
135 IFG(GP,GQ)< 2G(GP,GQ)=1
140 IFG(GP,GQ)>2G(GP,GQ)=3
145 NEXTGQ,GP
150 FORMI=ØTO4:FORMJ=ØTO4:GOSUB175
155 IFM(MI,MH,Ø)>1THEN165
160 M(MI,MJ,1)=G(GP,Ø):M(MI,MJ,2)=G(GP,1):M(MI,MJ,Ø)=1:POKE16327,
    42:POKE16327,32
165 NEXTMJ,MI
170 POKE16325,32:GF=Ø:RETURN
175 REM ** GAMMA PHASE-2 **
180 IFMI<2ANDMJ>2GP= Ø
182 IFMI<2ANDMJ=2GP=1
184 IFMI<2ANDMJ>2GP=2
186 IFMI=2ANDMJ<2GP=3
188 IFMI=2ANDMJ>2GP=5
190 IFM>2ANDMJ>2GP=5
192 IFMI>2ANDMJ=2GP=6
194 IFMI>2ANDMJ>2GP=7
196 RETURN
```

```
200 REM ** GAMMA CONL G-1 **
205 MI=I+2:MJ=J+2:RI=M(MI,MJ,1):RJ=M(MI,MJ,2):KL=M(MI,MJ,Ø):
    IFKL=ØTHEN235
210 I=RI-2:J=RJ-2:GOSUB725:IFCO=32THEN24Ø
215 C=C+1:IFKL=ØTHEN235
220 KL=KL-1:IFKL=ØTHEN235
225 IFKL=3THEN23ØELSE21Ø
230 GF=1:GOTO21Ø
235 GOSUB65Ø:KL=1:GOTO21Ø
240 IFKL<4THEN245ELSE255
245 KL=KL+1:IFKL=4THEN25ØELSE255
250 GF=1
255 M(MI,MJ,1)=RI:M(MI,MJ,2)=RJ:M(MI,MJ,Ø)=KL:D=D+1:RETURN
```

A WORD ABOUT GAMMA-IIG

Even using these program-compacting procedures, I am still finding it difficult to work a GAMMA-II program into a 4K computer format. I'm not saying it cannot be done, rather I haven't yet found a good way to go about it.

For the benefit of experimenters having systems with 16K or more memory, or for those who think they have some ideas about fitting the scheme into a 4K format, I have included a flowchart for GAMMA-IIG.

The flowchart does not call any subroutines that haven't been described elsewhere in this chapter, so there aren't any serious theoretical problems. All you have to do is write the master program yourself. Good luck—it's a great way to test your own understanding of everything you've done so far.

COMPOSITE PROGRAM LISTINGS

ALPHA-IG Composite Listing

```
10 REM ** ALPHA-iG MASTER **

15 CLS:PRINT@0,"*ALPHA-I*":GOSUB550:GOSUB600

20 GOSUB725:IFCO=32THEN35

25 C=C+1:GOSUB650:I=RI-2:J=RJ-2:GOSUB725:IFCO<>32THEN25

30 D=D+1:GOSUB800

35 GOSUB400:IFINKEY$="D"THEN10ELSE20

400 REM ** MOVE II-1 **

405 POKENP,32:NP=NP+I+64*J:POKENP,140:RETURN

550 REM ** FIELD II-1 **

555 FORF=15553T015615:POKEF,131:NEXT

560 FORF=16193T016255:POKEF,176:NEXT

565 FORF=15553T016255STEP64:POKEF,191:NEXT

570 FORF=15615T016255STEP64:POKEF,191:NEXT
```

```
575 RETURN

600 REM ** INITIAL IIG-1 **

605 NP=15968:GOSUB650:I=RI-2:J=RJ-2:C=0:D=0:RETURN

650 REM ** FETCH NEW IIG-1 **

655 RI=RND(5)-1:RJ=RND(5)-1:IFRI=2ANDRJ=2THEN655

660 RETURN

725 REM ** CON SENSE II-1 **

730 FORSI=ABS(I)TO1STEP-1:FORSJ=ABS(J)TO1STEP-1

735 CO=PEEK(NP+SI*SGN(I)+64*SJ*SGN(J))

740 IFCO<>32RETURN

745 NEXTSJ,SI:RETURN

800 REM ** UD SCORE IIG-1 **

810 PRINT@13,"NO. CONTACTS":PRINT@30,"NO. GOOD MOVES":PRINT
    @50,"SCORE"

815 PRINT@79,C:PRINT@99,D:PRINT@114,USING"#.##";D/C:PRINT@
    125,CHR$(C1)

820 RETURN
```

ALPHA-IIG Composite Listing

```
10 REM ** ALPHA-iIG MASTER **

15 CLS:PRINT@0,"*ALPHA-II*":PRINT@60,"FIG.":GOSUB550:GOSUB
   675:GOSUB600

20 GOSUB725:IFCO=32THEN35

25 C1=CO:C=C+1:GOSUB650:I=RI-2:J=RJ-2:GOSUB725:IFCO<>32THEN
   25

30 D=D+1:GOSUB800

35 GOSUB400:IFINKEY$="D"THEN10ELSE20

400 REM ** MOVE II-1 **

405 POKENP,32:NP=NP+I+64*J:POKENP,140:RETURN

550 REM ** FIELD II-1 **

555 FORF=15553TO15615:POKEF,131:NEXT

560 FORF=16193TO16255:POKEF,176:NEXT

565 FORF=15553TO16255STEP64:POKEF,191:NEXT
```

313

```
570 FORF=15615T016255STEP64:POKEF,191:NEXT

575 RETURN

600 REM ** INITIAL IIG-1 **

605 NP=15968:GOSUB650:I=RI-2:J=RJ-2:C=0:D=0:RETURN

650 REM ** FETCH NEW IIG-1 **

655 RI=RND(5)-1:RJ=RND(5)-1:IFRI=2ANDRJ=2THEN655

660 RETURN

675 REM ** FIGS II-1 **

677 P=15755

680 FORF=0TO1:FORN=P+64*FTOP+64*F+8:POKEN,35:NEXTN,F

682 FORF=0TO1:FORN=P+32+64*FTOP+40+64*F:POKEN,48:NEXTN,F

684 FORF=0TO1:FORN=P+192+64*FTOP+200+64*F:POKEN,42:NEXTN,F

685 FORF=0TO1:FORN=P+32+64*FTOP+40+64*F:POKEN,48:NEXTN,F

686 FORF=0TO1:FORN=P+224+64*FTOP+232+64*F:POKEN,37:NEXTN,F

690 RETURN

725 REM ** CON SENSE II-1 **

730 FORSI=ABS(I)TO1STEP-1:FORSJ=ABS(J)TO1STEP-1

735 CO=PEEK(NP+SI*SGN(I)+64*SJ*SGN(J))

740 IFCO<>32RETURN

745 NEXTSJ,SI:RETURN

800 REM ** UD SCORE IIG-1 **

810 PRINT@13,"NO. CONTACTS":PRINT@30,"NO. GOOD MOVES":PRINT
    @50,"SCORE"

815 PRINT@79,C:PRINT@99,D:PRINT@114,USING"#.##";D/C:PRINT@
    125,CHR$(C1)

820 RETURN
```

BETA-IG Composite Listing

```
10 REM ** BETA-IG MASTER **

15 DIMM(4,4,2)

20 CLS:PRINT@0,"*BETA-I*":GOSUB550:GOSUB600

25 GOSUB725:IFCO=32THEN45

30 C=C+1:MI=I+2:MJ=J+2:RI=M(MI,MJ,1):RJ=M(MI,MJ,2):KL=M(MI
   ,MJ,0):IFKL=0THEN50

35 I=RI-2:J=RJ-2:GOSUB725:IFCO<>32THEN55
```

314

```
40 D=D+1:GOSUB800

45 GOSUB400:IFINKEY$="D"THEN20ELSE25

50 GOSUB650:M(MI,MJ,1)=RI:M(MI,MJ,2)=RJ:M(MI,MJ,0)=1:GOTO35

55 C=C+1:GOTO50

400 REM ** MOVE II-1 **

405 POKENP,32:NP=NP+I+64*J:POKENP,140:RETURN

550 REM ** FIELD II-1 **

555 FORF=15553TO15615:POKEF,131:NEXT

560 FORF=16193TO16255:POKEF,176:NEXT

565 FORF=15553TO16255STEP64:POKEF,191:NEXT

570 FORF=15615TO16255STEP64:POKEF,191:NEXT

575 RETURN

600 REM ** INITIAL IIG-1 **

605 NP=15968:GOSUB650:I=RI-2:J=RJ-2:C=0:D=0:RETURN

650 REM ** FETCH NEW IIG-1 **

655 RI=RND(5)-1:RJ=RND(5)-1:IFRI=2ANDRJ=2THEN655

660 RETURN

725 REM ** CON SENSE II-1 **

730 FORSI=ABS(I)TO1STEP-1:FORSJ=ABS(J)TO1STEP-1

735 CO=PEEK(NP+SI*SGN(I)+64*SJ*SGN(J))

740 IFCO<>32RETURN

745 NEXTSJ,SI:RETURN

800 REM ** UD SCORE IIG-1 **

810 PRINT@13,"NO. CONTACTS":PRINT@30,"NO. GOOD MOVES":PRINT
    @50,"SCORE"

815 PRINT@79,C:PRINT@99,D:PRINT@114,USING"#.##";D/C:PRINT@
    125,CHR$(C1)

820 RETURN
```

BETA-IIG Composite Listing

```
10 REM ** BETA-IIG MASTER **

15 DIMM(4,4,4,2)

20 CLS:PRINT@0,"*BETA-II*":PRINT@60,"FIG.":GOSUB550:GOSUB
   675:GOSUB600

25 GOSUB725:IFCO=32THEN45
```

315

```
30 C=C+1:C1=CO:GOSUB900:MI=I+2:MJ=J+2:RI=M(MI,MJ,B,1):RJ=M
   (MI,MJ,B,2):IFM(MI, MJ,B,0)=0THEN50
35 I=RI-2:J=RJ-2:GOSUB725:IFCO<>32THEN55
40 D=D+1:GOSUB800
45 GOSUB400:IFINKEY$="D"THEN20ELSE25
50 GOSUB650:M(MI,MJ,B,1)=RI:M(MI,MJ,B,2)=2:M(MI,MJ,B,0)=1:
   GOTO35
55 C=C+1:GOTO50
400 REM ** MOVE II-1 **
405 POKENP,32:NP=NP+I+64*J:POKENP,140:RETURN
550 REM ** FIELD II-1 **
555 FORF=15553TO15615:POKEF,131:NEXT
560 FORF=16193TO16255:POKEF,176:NEXT
565 FORF=15553TO16255STEP64:POKEF,191:NEXT
570 FORF=15615TO16255STEP64:POKEF,191:NEXT
575 RETURN
600 REM ** INITIAL IIG-1 **
605 NP=15968:GOSUB650:I=RI-2:J=RJ-2:C=0:D=0:RETURN
650 REM ** FETCH NEW IIG-1 **
655 RI=RND(5)-1:RJ=RND(5)-1:IFRI=2ANDRJ=2THEN655
660 RETURN
675 REM ** FIGS II-1 **
677 P=15755
680 FORF=0TO1:FORN=P+64*FTOP+64*F+8:POKEN,35:NEXTN,F
682 FORF=0TO1:FORN=P+32+64*FTOP+40+64*F:POKEN,48:NEXTN,F
684 FORF=0TO1:FORN=P+192+64*FTOP+200+64*F:POKEN,42:NEXTN,F
685 FORF=0TO1:FORN=P+32+64*FTOP+40+64*F:POKEN,48:NEXTN,F
686 FORF=0TO1:FORN=P+224+64*FTOP+232+64*F:POKEN,37:NEXTN,F
690 RETURN
725 REM ** CON SENSE II-1 **
730 FORSI=ABS(I)TO1STEP-1:FORSJ=ABS(J)TO1STEP-1
735 CO=PEEK(NP+SI*SGN(I)+64*SJ*SGN(J))
740 IFCO<>32RETURN
745 NEXTSJ,SI:RETURN
```

```
800 REM ** UD SCORE IIG-1 **

810 PRINT@13,"NO. CONTACTS":PRINT@30,"NO. GOOD MOVES":PRINT
    @50,"SCORE"

815 PRINT@79,C:PRINT@99,D:PRINT@114,USING"#.##";D/C:PRINT@
    125,CHR$(C1)

820 RETURN

900 REM ** CON EVAL IIG-1 **

905 IFCO=131ORCO=176ORCO=191THEN930

910 IFCO=35THEN935

915 IFCO=48THEN940

920 IFCO=42THEN945

925 IFCO=37THEN950

930 B=0:RETURN

935 B=1:RETURN

940 B=2:RETURN

945 B=3:RETURN

950 B=4:RETURN
```

GAMMA-IG Composite Listing

```
10 REM ** GAMMA-IG MASTER **

15 DIMM(4,4,2):DIMG(7,1):DIMK(7)

20 CLS:PRINT@0,"*GAMMA-I*":GOSUB550:GOSUB600

25 GOSUB725:IFCO=32THEN35

30 C=C+1:GOSUB200:GOSUB800:IFGF=1THEN40

35 GOSUB400:IFINKEY$="D"THEN20ELSE25

40 GOSUB100:GOTO35

100 REM ** GAMMA-GEN G1 **

105 POKE16325,35:FORK=0TO7:K(K)=0:FORL=0TO1:G(K,L)=0:NEXTL,
    K:FORMI=0TO4:FORMJ=0TO4:GOSUB175:IFM(MI,MJ,0)<4THEN115

110 K(GP)=K(GP)+1:G(GP,0)=G(GP,0)+M(MI,MJ,1):G(GP,1)=G(GP,1
    )+M(MI,MJ,2)

115 NEXTMJ,MI

120 FORGP=0TO7:FORGQ=0TO1

125 IFK(GP)=0THEN145

130 G(GP,GQ)=G(GP,GQ)/K(GP)

135 IFG(GP,GQ)<2G(GP,GQ)=1
```

317

```
140 IFG(GP,GQ)>2G(GP,GQ)=3

145 NEXTGQ,GP

150 FORMI=0TO4:FORMJ=0TO4:GOSUB175

155 IFM(MI,MJ,0)>1THEN165

160 M(MI,MJ,1)=G(GP,0):M(MI,MJ,2)=G(GP,1):M(MI,MJ,0)=1:POKE
    16327,42:POKE16327,32
165 NEXTMJ,MI

170 POKE16325,32:GF=0:RETURN

175 REM ** GAMMA PHASE-2 **

180 IFMI<2ANDMJ<2GP=0

182 IFMI<2ANDMJ=2GP=1

184 IFMI<2ANDMJ>2GP=2

186 IFMI=2ANDMJ<2GP=3

188 IFMI=2ANDMJ>2GP=4

190 IFMI>2ANDMJ<2GP=5

192 IFMI>2ANDMJ=2GP=6

194 IFMI>2ANDMJ>2GP=7

196 RETURN

200 REM ** GAMMA CONL G-1 **

205 MI=I+2:MJ=J+2:RI=M(MI,MJ,1):RJ=M(MI,MJ,2):KL=M(MI,MJ,0
    ):IFKL=0THEN235
210 I=RI-2:J=RJ-2:GOSUB725:IFCO=32THEN240

215 C=C+1:IFKL=0THEN235

220 KL=KL-1:IFKL=0THEN235

225 IFKL=3THEN230ELSE210

230 GF=1:GOTO210

235 GOSUB650:KL=1:GOTO210

240 IFKL<4THEN245ELSE255

245 KL=KL+1:IFKL=4THEN250ELSE255

250 GF=1

255 M(MI,MJ,1)=RI:M(MI,MJ,2)=RJ:M(MI,MJ,0)=KL:D=D+1:RETURN

400 REM ** MOVE II-1 **

405 POKENP,32:NP=NP+I+64*J:POKENP,140:RETURN

550 REM ** FIELD II-1 **
```

```
555 FORF=15553T015615:POKEF,131:NEXT

560 FORF=16193T016255:POKEF,176:NEXT

565 FORF=15553T016255STEP64:POKEF,191:NEXT

570 FORF=15615T016255STEP64:POKEF,191:NEXT

575 RETURN

600 REM ** INITIAL IIG-1 **

605 NP=15968:GOSUB650:I=RI-2:J=RJ-2:C=0:D=0:RETURN

650 REM ** FETCH NEW IIG-1 **

655 RI=RND(5)-1:RJ=RND(5)-1:IFRI=2ANDRJ=2THEN655

660 RETURN

725 REM ** CON SENSE II-1 **

730 FORSI=ABS(I)TO1STEP-1:FORSJ=ABS(J)TO1STEP-1

735 CO=PEEK(NP+SI*SGN(I)+64*SJ*SGN(J))

740 IFCO<>32RETURN

745 NEXTSJ,SI:RETURN

800 REM ** UD SCORE IIG-1 **

810 PRINT@13,"NO. CONTACTS":PRINT@30,"NO. GOOD MOVES":PRINT
    @50,"SCORE"
815 PRINT@79,C:PRINT@99,D:PRINT@114,USING"#.##";D/C:PRINT
    @125,CHR$(C1)

820 RETURN
```

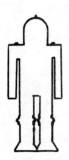

Appendix A
Standard ASCII Codes
And Characters

All keyboard characters are associated with a standard ASCII decimal number between 32 and 95 inclusively. There are two ways to display these characters on the screen of a TRS-80 system:

(1) Do a program statement or keyboard command of this form: PRINT CHR$(*n*), where *n* is the ASCII code number shown on the chart in Fig. A-1.

(2) Do a POKE*n,x* where *n* is the ASCII code number and *x* is the POKE graphics screen-position number (between 15360 and 16382.)

CODE	CHAR	CODE	CHAR	CODE	CHAR	CODE	CHAR
32	space	49	1	66	B	83	S
33	!	50	2	67	C	84	T
34	"	51	3	68	D	85	U
35	#	52	4	69	E	86	V
36	$	53	5	70	F	87	W
37	%	54	6	71	G	88	X
38	&	55	7	72	H	89	Y
39	'	56	8	73	I	90	Z
40	(	57	9	74	J	91	↑
41	)	58	:	75	K	92	↓
42	*	59	;	76	L	93	←
43	+	60	<	77	M	94	→
44	,	61	=	78	N	95	—
45	−	62	.	79	O		
46	.	63	?	80	P		
47	/	64	@	81	Q		
48	0	65	A	82	R		

Fig. A-1. ASC II codes and characters.

Appendix B
TRS-80 Graphics Codes And Figures

The TRS-80 employs a scheme for generating simple graphics figures. Each of the figures in Fig. A-2 has a number associated with it, and the figures are displayed in a fashion identical to that of displaying standard ASCII characters by means of the POKE statement.

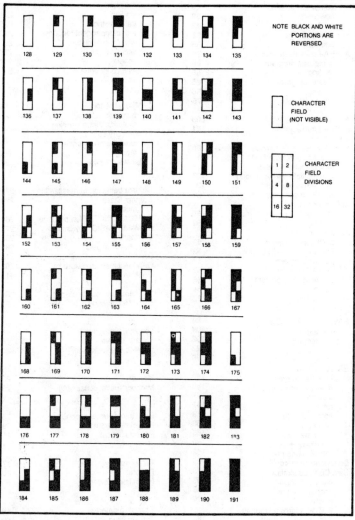

Fig. B-1. TRS-80 graphics.

Index